158 Topics in Current Chemistry

Photoinduced Electron Transfer II

Editor: J. Mattay

With contributions by
V. Balzani, F. Barigelletti, C. A. Bignozzi, R. Billing,
C. Chiorboli, L. De Cola, H. Hennig, M. T. Indelli,
H. Kunkely, D. Rehorek, F. Scandola, A. Vogler

With 35 Figures and 23 Tables

Springer-Verlag
Berlin Heidelberg New York London
Paris Tokyo Hong Kong Barcelona

This series presents critical reviews of the present position and future trends in modern chemical research. It is addressed to all research and industrial chemists who wish to keep abreast of advances in their subject.

As a rule, contributions are specially commissioned. The editors and publishers will, however, always be pleased to receive suggestions and supplementary information. Papers are accepted for "Topics in Current Chemistry" in English.

ISBN 3-540-52568-8 Springer-Verlag Berlin Heidelberg New York
ISBN 0-387-52568-8 Springer-Verlag New York Berlin Heidelberg

Typesetting: Th. Müntzer, Bad Langensalza; Printing: Heenemann, Berlin;

Bookbinding: Lüderitz & Bauer, Berlin
2151/3020-543210 — Printed on acid-freepaper

Guest Editor

Prof. Dr. *Jochen Mattay*
Organisch-Chemisches Institut,
Westfälische Wilhelms-Universität Münster,
Orléansring 23, D-4400 Münster

Editorial Board

Preface to the Series on Photoinduced Electron Transfer

The exchange of an electron from a donor molecule to an acceptor molecule belongs to the most fundamental processes in artificial and natural systems, although, at the primary stage, bonds are neither broken nor formed. However, the transfer of an electron determines the chemical fate of the molecular entities to a great extent. Nature has made use of this principle since the early beginnings of life by converting light energy into chemical energy via charge separation. In recent years, man has learnt, e.g. from X-ray analyses performed by Huber, Michel and Deisenhofer, how elaborately the molecular entities are constructed within the supermolecular framework of proteins. The light energy is transferred along cascades of donor and acceptor substrates in order to prevent back electron transfer as an energy wasting step and chemical changes are thus induced in the desired manner.

Today we are still far from a complete understanding of light-driven electron transfer processes in natural systems. It is not without reason that the Pimentel Report emphasizes the necessity of future efforts in this field, since to understand and "to replicate photosynthesis in the laboratory would clearly be a major triumph with dramatic implications". Despite the fact that we are at the very beginning of knowledge about these fundamental natural processes, we have made much progress in understanding electron transfer reactions in "simple" molecular systems. For example, most recently, a unified view of organic and inorganic reaction mechanisms has been discussed by Kochi. In this context, photochemistry plays a crucial role not only for the reasons mentioned above, but also as a tool to achieve electron transfer reactions. The literature contains a host of examples, inorganic as well as organic, homogeneous as well as heterogeneous. Not surprisingly, most of them have been published within the last decade, although early examples have been known since the beginning of photochemistry (cf. Roth's article). A reason is certainly the rapid development of analytical methods, which makes possible the study of chemical processes at very short time ranges. Eberson in his monograph, printed by this publishing company two years ago, nicely pointed out that "electron transfer theories come in cycles". Though electron transfer has been known to inorganic chemists for a relatively long time, organic chemists have still to make up for missing concepts (cf. Eberson).

A major challenge for research in the future, the "control of chemical reactions" as stated by the Pimentel Report, can be approached by various methods; light-driven processes are among the most important ones. Without interaction of the diverse scientific disciplines, recent progress in photochemistry, as well as future developments would scarcely be possible. This is particularily true for the study of electron transfer processes. In this context lies a challenge for science and economy and the special fascination of this topic — at least for the guest editor.

The scope of photochemistry and the knowledge about the fundamentals of photoinduced electron transfer reactions have tremendously broadened over the last decade, as have their applications. Therefore I deeply appreciate that the Springer-Verlag has shown interest in this important development and is introducing a series of volumes on new trends in this field. It is clear that not all aspects of this rapidly developing topic can be exhaustively compiled. I have therefore tried to select some papers which most representatively reflect the current state of research. Several important contributions might be considered missing by those readers who are currently involved in this field, however, these scientists are referred to other monographs and periodical review series which have been published recently. These volumes are meant to give an impression of this newly discovered reaction type, its potential and on the other hand to complement other series.

The guest editor deeply appreciates that well-known experts have decided to contribute to this series. Their effort was substantial and I am thankful to all of them. Finally, I wish to express my appreciation to Dr. Stumpe and his coworkers at the Springer-Verlag for helping me with all the problems which arose during the process of bringing the manuscript together.

Münster, December 1989 Jochen Mattay

Preface to Volume II

The second volume of the PET (Photoinduced Electron Transfer) series in "Topics of Current Chemistry" is concerned with inorganic chemistry. The guest editor is very pleased that authors from four active and well-known groups have contributed to this volume. The articles are arranged according to their date of submission.

The photochemistry of transition metal complexes induced by outer sphere charge transfer excitation is the topic of the first contribution. Various types of aggregates, for example neutral donor-acceptor species or ion pairs, composed of two complexes or of a complex and an organic counter ion, are discussed with respect to their photophysical and photochemical behaviour.

Transition metal complexes which can be used as mediators for photochemical and chemiluminescent processes are dealt with in the second article. They are important for theoretical reasons as well as for applications such as water splitting, artificial photosynthesis, photogalvanic effects and their reversal, electrochemiluminescence etc.

A different field is dealt with in the third chapter, which is devoted to energy and electron transfer in polynuclear coordination compounds. Due to the large moleculares entities, containing several transition metal subunits, the properties of these "supermolecules" turn out to be complex, offering new interesting applications in the future (e.g. photonic molecular devices, capable of elaborating light signals).

A special type of supermolecular interaction is discussed in the final chapter. Pairing of oppositely charged ions be utilized to facilitate electron transfer processes after light absorption. The requirements for ion pair formation and the resulting influence on spectroscopic and photochemical properties is discussed in detail.

As shown in the various contributions of this volume, we are at the very beginning of a new and exciting field of inorganic photochemistry. New mechanistic aspects and unconventional applications will surely evolve in the future.

Münster, February 1990 Jochen Mattay

Table of Contents

Table of Contents of Volume 156

Photoinduced Electron Transfer I

Photochemistry of Transition Metal Complexes Induced by Outer-Sphere Charge Transfer Excitation

Arnd Vogler and Horst Kunkely

Institut für Anorganische Chemie, Universität Regensburg,
Universitätsstraße 31, D-8400 Regensburg

Table of Contents

The intermolecular (outer sphere, OS) interaction of a reducing and an oxidizing metal complex generates a new optical transition involving charge transfer (CT) from the electron donor to the acceptor. OS CT transitions are classified according to the redox site (metal or ligand). Generally, the interaction between donor and acceptor is facilitated by ion pairs consisting of an oxidizing complex cation and a reducing complex anion. There are also ion pairs which are composed of a metal complex and an organic counter ion as electron donor or acceptor. In addition, the review

Topics in Current Chemistry, Vol. 158

includes examples of OS CT interaction which do not involve ion pairs at all. — A short introduction into the theory is followed by the discussion of the spectroscopy of OS CT of transition metal complexes. Finally, photoreactions induced by OS CT transitions are reviewed. The optical transfer is frequently followed by a rapid back electron transfer which regenerates the starting complexes. In many cases the primary products are kinetically labile and substitution reactions compete successfully with back electron transfer. As a result stable redox products may be formed. As an alternative, the substitution can be followed by back electron transfer. Product formation appears then as a substitution of the starting complexes. The various possibilities are illustrated by appropriate examples.

1 Introduction

The exchange of an electron between two molecules may be considered to be the most fundamental and important chemical reaction. Such a redox process can occur thermally or photochemically. Intermolecular light-induced electron transfer involving transition metal complexes has been extensively studied during the last 15 years [1–8]. This interest was stimulated, at least partially, by attempts to develop an artificial photosynthesis for the conversion and chemical storage of solar energy [9–11]. It is well known that natural photosynthesis requires a light-induced electron transfer as the basic process.

Intermolecular photochemical electron transfer takes place by two different mechanisms. First, an electronically excited molecule may undergo an electron exchange with another molecule in its ground state. Secondly, a direct optical electron transfer can be achieved if the electron donor and acceptor are electronically coupled by a close contact. The majority of studies involving coordination compounds has been devoted to excited state electron transfer [1–8] while much less information is available on the second mechanism [6–8, 12–15] which is the subject of the present review.

Intermolecular and outer sphere (OS) electron transfer are frequently used as synonymous expressions. Generally, this is justified. However, in some cases both terms are not equivalent. While an intermolecular electron transfer is indeed always of the OS type, an OS electron transfer is not necessarily an intermolecular process. For example, in the complex $[(NH_3)_5Ru^{II}SC_6H_{12}SRu^{III}(NH_3)_5]^{5+}$ an optical electron transfer from Ru^{II} to Ru^{III} takes place [16]. This intramolecular process occurs by an OS mechanism since the electronic coupling between both metals is not mediated by connecting atoms. In distinction to an inner sphere (IS) transfer the electronic interaction does not take place via chemical bonds but rather through space. However, for most practical purposes intermolecular and OS electron transfer describe equivalent processes.

While the electronic coupling by an OS interaction is generally rather weak it may become quite strong if it is of the IS type although there is also a number of remarkable exceptions [17]. An IS mechanism is always in operation when the donor and acceptor site are connected by a direct bond without an intervening atom. Before we enter the discussion of light-induced OS electron transfer it is appropriate to mention briefly the typical optical IS charge transfer (CT) transitions of metal complexes which are classified according to the redox sites [18].

Ligand to metal CT (LMCT) absorption bands appear at long wavelength if the ligand is reducing and the metal oxidizing. Fe(III) and Co(III) complexes are well-documented examples. LMCT bands cause the colors of d^0 oxometallates such as CrO_4^{2-} (yellow) and MnO_4^- (violet).

Metal to ligand (MLCT) is another classical optical transition of metal complexes. MLCT absorptions are observed at long wavelength if the metal is reducing and a ligand provides low-energy empty orbitals. Complexes such as $[Fe(CN)_6]^{4-}$ and $[Ru(bipy)_3]^{2+}$ (bipy = 2,2′-bipyridyl) are typical cases. In addition, organometallics which contain a metal in a low oxidation state and

π-accepting ligands such as an olefin or an aromatic molecule are characterized by low-energy MLCT bands [18, 19].

Metal to metal CT (MMCT) bands occur only in the absorption spectra of binuclear (or polynuclear) complexes which contain a reducing and an oxidizing metal. Two cases can be distinguished. Both metals are either bridged by a suitable ligand (M_{red}—L—M_{ox}) [2, 12, 17–24] or connected by a direct, but polar metal-metal bond (M_{red}—M_{ox}). The binuclear complexes $[(NH_3)_5Co^{III}$-$NCRu^{II}(CN)_5]^-$ [12, 17] and $[Ph_3PAu^I$—$Co^{-I}(CO)_4]$ [25] are typical compounds which show low-energy MMCT bands.

Ligand to ligand CT (LLCT) absorptions were identified only recently [26]. These bands appear if one ligand is reducing and another oxidizing (L_{red}—M—L_{ox}). In simple cases L_{red} may be a halide, thiolate, aryl or alkyl group while polypyridyls such as bipy serve as oxidizing ligands. A specific form of L_{red}—M—L_{ox} complexes which display LLCT absorptions are ligand-based mixed-valence complexes. These compounds contain the same ligand in an oxidized and reduced form.

Let us now return to optical OS CT of metal complexes. The following discussion is presented in several sections. First, a brief description of the theoretical background is given. Secondly, the spectroscopy is discussed. Finally, the photochemistry induced by optical OS CT of metal complexes is treated.

2 Theoretical Background

An optical OS CT transition may occur if a reducing and an oxidizing molecule or ion are in close contact which provides some orbital overlap of the donor and acceptor. This close contact is frequently facilitated by the electrostatic attraction within an ion pair. But also neutral molecules may be close enough under suitable conditions, particularly at high concentrations or in the solid state. The electronic interaction between an electron donor and acceptor and the resulting optical CT transition can be understood on the basis of a theory advanced by N. S. Hush [21, 22, 24, 27, 28]. Although this theory was first developed for donating and accepting metal centers it can be applied to any other redox site as well. The Hush model in its simplest form requires a weak electronic interaction between donor and acceptor. This is certainly valid for the majority of the OS systems discussed in this article since donor and acceptor are not coupled by chemical bonds. The overlap of the orbitals involved in the OS CT transition is assumed to be rather small. The electronic spectrum of the donor/acceptor pair consists then of the superimposed spectra of the single components. In addition, a new absorption band appears which belongs to the optical CT transition from the donor (D) to the acceptor (A).

The energy of this OS CT transition (E_{CT}) depends on the potential difference ΔE between the redox couples D/D^+ and A/A^- and on the reorganizational energy χ (Fig. 1).

$$E_{CT} = \Delta E + \chi$$

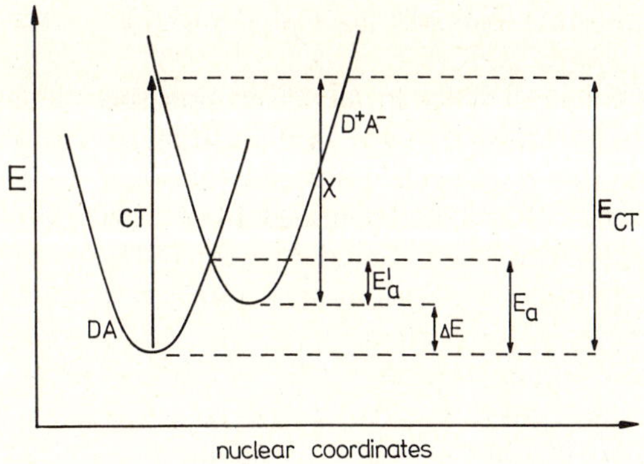

Fig. 1. Potential energy diagram for electron transfer from a donor (D) to an acceptor (A)

The parameter χ consists of an outer and an inner part.

$$\chi = \chi_0 + \chi_i$$

The inner contribution χ_i is a fraction of the Franck-Condon CT transition as shown in Fig. 1. It depends on the structural distortion which accompanies electron transfer. In the case of a metal complex this structural reorganization which may be associated with changes of the metal-ligand bond length varies with the oxidation state of the metal. Frequently, reduction is associated with an extension of the metal-ligand distance when an antibonding orbital is populated.

The optical CT as a Franck-Condon transition terminates in a vibrationally excited state of the redox isomer D^+A^- before it relaxes. The electron transfer can not only be achieved by light absorption but also as a thermal process which requires the activation energy E_{th} to reach the crossing point of both potential curves (Fig. 1). When the redox isomer has relaxed to its vibrationally ground state it may undergo a thermal back electron transfer by overcoming the activation energy $E'_a = E_{th} - \Delta E$.

While χ_i is an intrinsic property of the redox pair AD the outer part χ_0 depends on the reorganization of the solvent environment.

$$\chi_0 = e^2\left(\frac{1}{2a_1} + \frac{1}{2a_1} - \frac{1}{d}\right)\left(\frac{1}{n^2} - \frac{1}{D}\right)$$

The parameters a_1 and a_2 are the radii of the donor and acceptor assuming spherical structures. This assumption seems to be justified for tetrahedral or octahedral complexes. However, in the case of planar electron donor or acceptors such a simple picture certainly does not apply. A further parameter is the distance d between D and A with $d = a_1 + a_2$ as the closest possible approach. The polarity of solvent contributes also to χ_0. This polarity is represented by the term $1/n^2 - 1/D$ and determined by the refractive index n and the static dielectric constant D of the solvent. It follows that with an increasing solvent polarity and an increasing distance between donor and acceptor also the term χ_0 and finally the energy of the optical CT transition become larger [29].

The outer contribution χ_0 to the overall reorganizational energy introduces a serious complication to the evaluation of an optical OS CT. A change of the solvent does not only effect χ_0 by a variation of the solvent polarity. It can also affect the mutual orientation of donor and acceptor, particularly the distance d. For example, when the donor and acceptor are ions an increasing solvent polarity may increase also the distance d by an extension of the solvation shell of the ions. An empirical correlation which apparently takes care of these complications has been recently introduced to evaluate and predict the energy of optical OS CT transitions. This increment system developed by Hennig, Bendix, and Billing works rather well [15, 30, 31]. If follows that the solvent polarity and the distance d seem to vary in a predictable fashion.

Light absorption into the OS CT band is a photoredox process by definition. Generally, the generation of D^+A^- is followed by a rapid back electron transfer which requires a rather small activation energy E_a' (Fig. 1). As a consequence a permanent chemical change does not take place. An irreversible formation of stable photoproducts can be only achieved if the redox isomer D^+A^- is able to undergo some further structural rearrangements. These secondary processes must be fast enough to compete with back electron transfer. For example, photoactivity is expected if $[Co(NH_3)_6]^{3+}$ is the electron acceptor. Upon reduction $[Co(NH_3)_6]^{2+}$ is generated. It is kinetically very labile and undergoes a rapid decomposition in aqueous solution [32]. When the product-forming step is not very fast back electron transfer is certainly favored but a cage escape of the primary redox pair D^+A^- may facilitate a secondary reaction.

3 Spectroscopy

In analogy to IS CT transitions [18] optical OS CT can be classified according to the predominant localization of donor and acceptor orbitals at the ligand or metal. In suitable cases OS MLCT, OS LMCT, OS LLCT, and OS MMCT transitions will then be observed. In addition, the donor or acceptor may not be a metal complex at all. The corresponding OS CT transitions are now of the complex to acceptor and donor to complex type. Again, donor and acceptor orbitals can be located at the metal or ligand.

The charge of the donor and acceptor is used as a further classification of OS CT. As discussed below the majority of OS CT is observed for ion pairs which consist of an accepting cation and a donating anion. It is quite understandable that the electronic interaction between donor and acceptor is facilitated by electrostatic attraction in the ion pairs. In systems which do not consist of ion pairs high concentrations of at least one redox partner is required. OS CT transitions of this type can be identified either in the solid state or if the donor or acceptor is the solvent. The latter transition is well known as CT to solvent (CTTS) transition [8, 19, 33]. The reverse process, namely optical CT from the solvent, has not yet been observed to our knowledge.

3.1 Ion Pairs

The intensity of an OS CT absorption of an ion pair is influenced by the solvent in various ways. More polar solvents will favor the dissociation of the ion pair. In less polar solvents the solvation of ions and thus the distance between donor and acceptor are much smaller. The orbital overlap and consequently the intensity of the OS CT bands of these contact ion pairs are expected to be much larger than those of solvent-separated or dissociated ion pairs. It follows that the detection of an OS CT band of ion pairs is often facilitated by the use of nonpolar solvents. Unfortunately, this choice is frequently limited by the low solubilities of the ion pairs in solvents of low polarity.

Occasionally it is questionable if observed CT absorptions are really due to ion pairs. In some cases they rather belong to ligand-bridged bi- or polynuclear complexes and are of IS CT type which are not discussed in this review. Caution must be applied if one complex ion provides ligands which are potentially bridging (e.g. CN^-) and the counter ion is kinetically not inert (e.g. $[Fe(H_2O)_6]^{3+}$ or $[Cu(H_2O)_6]^{2+}$). Under these conditions it is likely that the donor and acceptor site are bridged by a ligand which mediates as IS CT interaction.

Another interesting but unresolved problem concerns the sign of the charge of the donor and acceptor ion. To our knowledge all studies have only dealt with ion pairs which consist of anionic donors and cationic acceptors but never with reversed ion pairs (cationic donors and anionic acceptors). This situation may be accidental since there does not seem to be any explanation of this observation. Ion pairs such as $Ag^+CrO_4^{2-}$ or $Ag^+MnO_4^-$ which display Ag^+ to Cr^{VI} or Mn^{VII} MMCT bands could be exceptions [34].

However, these CT bands were only detected for the salts in the solid state. Again, it is not sure if these salts can be really considered as ion pairs or if donor and acceptor interact via bridging ligands (e.g. Ag—O—Cr).

3.1.1 Complex to Complex Charge Transfer

MLCT

Aqueous solutions of the ion pairs $[Rh(bipy)_3]^{3+}[M(CN)_6]^{4-}$ with M = Fe, Ru, and Os display M^{II} to bipy OS MLCT absorptions at λ_{max} = 480 nm (ε = 61) for M = Fe, 400 nm (155) for Os, and 379 nm (110) for Ru [35]. As expected the OS CT bands shift to lower energies in the order of increasing reducing strength of $[M(CN)_6]^{4-}$ ($E_{1/2}$ = 0.19 V for Fe, 0.40 V for Os and 0.70 V vs SCE for Ru). The reorganizational energy which is associated with electron transfer was estimated to be ~15000 cm^{-1} for all three ion pairs.

LMCT

The electronic spectrum of the aqueous ion pair $[Ru(NH_3)_6]^{3+}[Rh(CN)_6]^{3-}$ contains a new absorption band at λ_{max} = 297 nm (ε = 29) which appears well resolved in the difference spectrum [36]. This band was assigned to an OS LMCT transition from cyanide coordinated at Rh^{III} to Ru^{III} of the counter ion. The

Table 1. Optical OS CT (metal to metal) Transitions of ion pairs

Acceptor	Donor	Abs. λ_{max}/nm	Solvent	Ref]
$[Co(NH_3)_6{}^{3+}$	$[Fe(CN)_6]^{4-}$	440	H_2O	[40]
$[Co(ethylenediamine)_3]^{3+}$	$[Fe(CN)_6]^{4-}$	430	H_2O	[40]
$[Co(NH_3)_6]^{3+}$	$[Ru(CN)_6]^{4-}$	360	DMSO	[41]
$[Co(NH_3)_6]^{3+}$	$[Ru(CN)_6]^{4-}$	344	H_2O	[42]
$[Co(ethylenediamine)_3]^{3+}$	$[Ru(CN)_6]^{4-}$	351	H_2O	[42]
$[Co(1,2\text{-diaminopropane})_3]^{3+}$	$[Ru(CN)_6]^{4-}$	353	H_2O	[42]
$[Co(1,2\text{-cyclohexanediamine})_3]^{3+}$	$[Ru(CN)_6]^{4-}$	359	H_2O	[42]
$[Co(sepulchrate)]^{3+}$	$[Ru(CN)_6]^{4-}$	374	H_2O	[42]
$[Ru(NH_3)_5(4\text{-bromopyridine})]^{3+}$	$[Fe(CN)_6]^{4-}$	932	H_2O	[46]
$[Ru(NH_3)_5(4\text{-chloropyridine})]^{3+}$	$[Fe(CN)_6]^{4-}$	940	H_2O	[46]
$[Ru(NH_3)_5(pyridine)]^{3+}$	$[Fe(CN)_6]^{4-}$	910	H_2O	[44, 46]
$[Ru(NH_3)_5(4\text{-methylpyridine})]^{3+}$	$[Fe(CN)_6]^{4-}$	898	H_2O	[46]
$[Ru(NH_3)_5(4\text{-t-butylpyridine}]^{3+}$	$[Fe(CN)_6]^{4-}$	894	H_2O	[46]
$[Ru(NH_3)_5(pyrazole)]^{3+}$	$[Fe(CN)_6]^{4-}$	864	H_2O	[46]
$[Ru(NH_3)_5(2,5\text{-dimethylpyrazole})]^{3+}$	$[Fe(CN)_6]^{4-}$	810	H_2O	[46]
$[Ru(NH_3)_5(imidazole)]^{3+}$	$[Fe(CN)_6]^{4-}$	788	H_2O	[46]
$[Ru(NH_3)_6]^{3+}$	$[Fe(CN)_6]^{4-}$	737	Nujol	[44]
$[Ru(NH_3)_6]^{3+}$	$[Fe(CN)_6]^{4-}$	714	H_2O	[45]
$[Ru(NH_3)_6]^{3+}$	$[Fe(CN)_5CO]^{3-}$	450	H_2O	[45]
$[Ru(NH_3)_6]^{3+}$	$[Fe(CN)_5(dimethylsulfoxide)]^{3-}$	558	H_2O	[45]
$[Ru(NH_3)_6]^{3+}$	$[Fe(CN)_5(pyrazine)]^{3-}$	667	H_2O	[45]
$[Ru(NH_3)_6]^{3+}$	$[Fe(CN)_5(pyridine)]^{3-}$	725	H_2O	[45]
$[Ru(NH_3)_6]^{3+}$	$[Fe(N)_5(imidazole)]^{3-}$	741	H_2O	[45]
$[Ru(NH_3)_6]^{3+}$	$[Ru(CN)_6]^{4-}$	541	Nujol	[44]
$[Ru(NH_3)_6]^{3+}$	$[Ru(CN)_6]^{4-}$	549	H_2O	[45]
$[Ru(NH_3)_5(3,5\text{-dimethylpyrazine})]^{3+}$	$[Ru(CN)_6]^{4-}$	700	H_2O	[46]
$[Ru(NH_3)_5(3,5\text{-dichloropyridine})]^{3+}$	$[Ru(CN)_6]^{4-}$	675	H_2O	[46]
$[Ru(NH_3)_5(\text{-chloropyridine})]^{3+}$	$[Ru(CN)_6]^{4-}$	665	H_2O	[46]
$[Ru(NH_3)_5(4\text{-bromopyridine})]^{3+}$	$[Ru(CN)_6]^{4-}$	653	H_2O	[46]
$[Ru(NH_3)_5(4\text{-chloropyridine})]^{3+}$	$[Ru(CN)_6]^{4-}$	656	H_2O	[46]

$[Ru(NH_3)_5(pyridine)]^{3+}$	$[Ru(CN)_6]^{4-}$	643	H_2O	[46]
$[Ru(NH_3)_5(4\text{-methylpyridine})]^{3+}$	$[Ru(CN)_6]^{4-}$	627	H_2O	[46]
$[Ru(NH_3)_5Cl]^{2+}$	$[Ru(CN)_6]^{4-}$	510	H_2O	[47]
$[Ru(NH_3)_5(3,5\text{-dimethylpyrazine})]^{3+}$	$[Os(CN)_6]^{4-}$	716	H_2O	[46]
$[Ru(NH_3)_5(3,5\text{-dichlorpyridine})]^{3+}$	$[Os(CN)_6]^{4-}$	700	H_2O	[46]
$[Ru(NH_3)_5(4\text{-bromopyridine})]^{3+}$	$[Os(CN)_6]^{4-}$	670	H_2O	[46]
$[Ru(NH_3)_5(3\text{-chlorpyridine})]^{3+}$	$[Os(CN)_6]^{4-}$	668	H_2O	[46]
$[Ru(NH_3)_5(pyridine)]^{3+}$	$[Os(CN)_6]^{4-}$	658	H_2O	[46]
$[Ru(NH_3)_5(4\text{-t.-butylpyridine})]^{3+}$	$[Os(CN)_6]^{4-}$	625	H_2O	[46]
$[Ru(NH_3)_5(4\text{-methylpyridine})]^{3+}$	$[Os(CN)_6]^{4-}$	626	H_2O	[46]
$[Os(NH_3)_5Cl]^{2+}$	$[Fe(CN)_6]^{4-}$	438	H_2O	[12, 42]
$[Os(NH_3)_5Cl]^{2+}$	$[Ru(CN)_6]^{4-}$	372	H_2O	[12, 42]
$[Os(NH_3)_5Cl]^{2+}$	$[Os(CN)_6]^{4-}$	388	H_2O	[12, 42]
$[Eu(2.2.1\ cryptand)]^{3+}$	$[Fe(CN)_6]^{4-}$	530	H_2O	[48, 49]
$[Eu(2.2.1\ cryptand)]^{3+}$	$[Ru(CN)_6]^{4-}$	434	H_2O	[48, 49]
$[Eu(2.2.1\ cryptand)]^{3+}$	$[Os(CN)_6]^{4-}$	450	H_2O	[48, 49]
$[Pt(NH_3)_5Cl]^{3+}$	$[Fe(CN)_6]^{4-}$	418	H_2O	[50]
$[Pt(NH_3)_5Cl]^{3+}$	$[Ru(CN)_6]^{4-}$	332	H_2O	[50]
$[Pt(NH_3)_5Cl]^{3+}$	$[Os(CN)_6]^{4-}$	353	H_2O	[50]
$[Pt(NH_3)_5Cl]^{3+}$	$[Pt(CN)_4]^{4-}$	295	H_2O	[50]

corresponding CN^- to Ru^{III} IS LMCT band of $[Ru(CN)_6]^{3-}$ is shifted to longer wavelength (λ_{max} = 475 nm). This shift is ascribed to the different redox potentials for the reduction of Ru^{III} ($E_{1/2}$ = −0.18 for $[Ru(NH_3)_6]^{3+}$ and +0.70 V for $[Ru(CN)_6]^{3-}$). In addition, the larger distance between CN^- and Ru^{III} in the ion pair contributes certainly to the higher energy of the OS LMCT band.

LLCT

Intense absorption bands which are assigned to IS LLCT transitions appear in the electronic spectra of square planar Ni^{II}, Pd^{II}, and Pt^{II} complexes which contain a 1,2-ethylenedithiolate as electron-donating and 1,2-diimine as accepting ligand [26]:

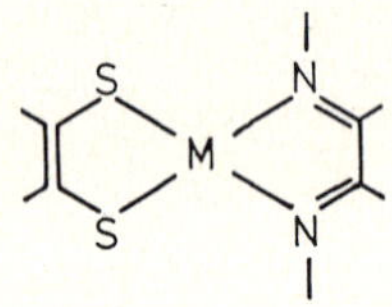

It the diimine and the dithiolate are coordinated in separate complexes which form an ion pair it should be possible to identify OS LLCT absorptions. Such bands were indeed detected in the spectra of the insoluble salts $[Ni(tim)]^{2+}[M(mnt)_2]^{2-}$ with M^{II} = Ni^{II}, Pd^{II}, and Pt^{II} at λ_{max} = 840 nm for Ni, 834 nm for Pd, and 824 nm for Pt [37].

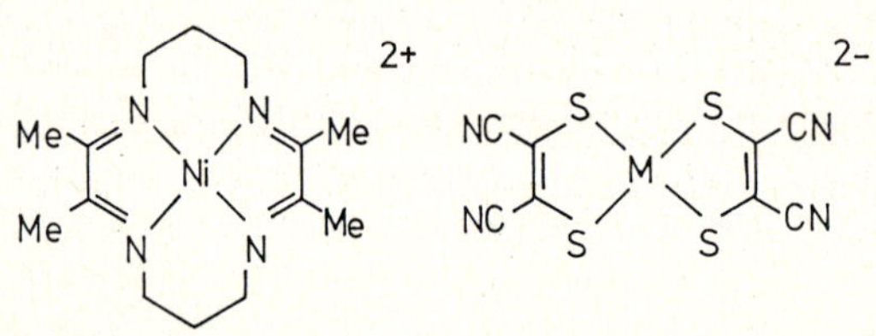

The intense and broad absorptions ($\varepsilon \sim 10^4$) at about 830 nm were assigned to OS LLCT transitions from mnt^{2-} to tim. This assignment is supported by the fact that the energy of this transition is almost independent of the metal M^{II}.

MMCT

The majority of OS CT transitions of ion pairs which consist only of complex ions is of the MMCT type. The combination of the oxidizing aquo cations Fe^{III}, Cu^{II}, UO_2^{2+} and VO^{2+} with the reducing cyano complex anions $[Fe^{II}(CN)_6]^{4-}$, $[Fe^{II}(CN)_5L]^{3-}$, $[Ru^{II}(CN)_6]^{4-}$, $[Mo^{IV}(CN)_8]^{4-}$ and $[W^{IV}(CN)_8]^{4-}$ causes also the appearance of new MMCT absorptions [38, 39]. However, it is questionable if they are OS in character. Since the aquo cations are kinetically labile and the cyano complex provide bridging ligands bi- or polynuclear complexes with an IS MMCT interaction may have been formed. For several systems this suspicion was confirmed [38, 39].

Cationic Co^{III} amine complexes are well suited as electron acceptors. When they are combined with the reducing anions $[Fe(CN)_6]^{4-}$ [40] or $[Ru(CN)_6]^{4-}$ [12, 41, 42] ion pairs are formed which are characterized by OS MMCT absorption bands (Table 1). The rather short wavelength of these absorptions is certainly due to the large reorganizational energy which is associated with the reduction of low-spin Co^{III}. Since an electron is accepted into an antibonding e_g orbital (in O_h symmetry) the MMCT transition requires a large extension of the cobalt-ligand bond distance.

It is quite interesting to compare the MMCT transition for the ion pair (OS) $[Co^{III}(NH_3)_6]^{3+}[Ru^{II}(CN)_6]^{4-}$ (λ_{max} = 344 nm) [12, 41, 42] and the related binuclear complex (IS) $[(NH_3)_5Co^{III}NCRu^{II}(CN)_5]^-$ (λ_{max} = 375 nm) [12, 43]. The blue shift for the ion pair is probably caused — at least partially — by the larger distance between the metal centers as redox sites.

A distance effect was also expected for the ion pairs $[CoL_6]^{3+}[Ru(CN)_6]^{4-}$ with $[Co(N{-}N)_3]^{3+}$ with N—N = ethylenediamine, 1,2-diaminopropane, and 1,2-cyclohexanediamine since the ligands become larger in this series [12, 42]. However, contrary to the expectation the OS MMCT bands were shifted to longer wavelength (Table 1). The distance between donor and acceptor grew apparently. It is assumed that the increasing size of the complex cations is indeed associated with a decreasing size of the solvated cobalt complexes in the ion pair. This phenomenon is well known for alkali cations. It is certainly also favored by the hydrophobicity of the Co complexes which becomes larger when the number of alkyl substituents at the diamine ligand increases.

The largest number of ion pairs which display OS MMCT bands was observed with octahedral ammine complexes of Ru^{III} as acceptor and $[Fe^{II}(CN)_6]^{4-}$ [44–46] $[Fe^{II}(CN)_5L]^{3-}$ [45], $[Ru(CN)_6]^{4-}$ [44–47], and $[Os(CN)_6]^{4-}$ [46] as donor (Table 1). Since Ru^{III} is a d^5 metal the lowest-energy MMCT transition terminates in the hole of the t_{2g} orbitals. Due to the non-bonding character of these orbitals the reorganizational energy is obviously much smaller than for low-spin Co^{III}. In fact, the reorganizational energy of all these Ru^{III}/Fe^{II}, Ru^{II}, and Os^{II} ion pairs seems to be rather similar. This assumption is supported by the observation that the energy of the MMCT transition is proportional to the redox asymmetry ΔE [45, 46]. For a homonuclear ion pair ΔE is generally rather small but can become rather large in suitable cases. It is as large as 0.90 V for $[Ru^{III}(NH_3)_5Cl]^{2+}[Ru^{II}(CN)_6]^{4-}$ since the NH_3 and Cl^- ligands stabilize the oxidation state III while CN^- as a π-accepting ligand favors Ru^{II} [47]. For many of these ion pairs it has been confirmed that electronic coupling between donor and acceptor is indeed rather small.

The ion pairs $[Os^{III}(NH_3)_5Cl]^{2+}[M^{II}(CN)_6]^{4-}$ (M = Fe, Ru, Os) display also long-wavelength OS MMCT bands [12, 42] (Table 1). Since $[Os(NH_3)_5Cl]^{2+}$ is much less oxidizing (—1.10 V) than $[Ru(NH_3)_5Cl]^{2+}$ (—0.04 V) the MMCT absorptions of the ion pairs which contain Os^{III} as acceptor appear at shorter wavelength than those of Ru^{III}.

Upon addition of aqueous solutions of Eu^{3+} to $[M(CN)_6]^{4-}$ with M = Fe, Ru, and Os insoluble salts precipitate. If Eu^{3+} is incorporated into a cryptand (C 2.2.1 = 4,7,13,16,21-pentaoxo-1,10-diazabicyclo-[8.8.5]tricosane) the cationic cryptate $[Eu^{III}C2.2.1]^{3+}$ forms soluble ion pairs with $[M(CN)_6]^{4-}$. These pairs are characterized by M^{II} to Eu^{III} OS MMCT absorptions [14, 48, 49] (Table 1). The MMCT state (Eu^{II}/M^{III}) can be also generated by excited state electron transfer. That portion of $[EuC2.2.1)^{3+}$ which is not ion-paired is luminescent but the emission is quenched by electron transfer in diffusional encounters.

The ion pairs $[Pt^{IV}(NH_3)_5Cl]^{3+}[M^{II}(CN)_6]^{4-}$ (M = Fe, Ru, Os) show M^{II} to Pt^{IV} OS MMCT bands at rather short wavelength [50] (Table 1) although the Pt^{IV} complex is an oxidant of moderate strength. However, the potentials of Pt^{IV} complexes are generally known only for two-electron reductions to Pt^{II} while the optical MMCT transition as a one-electron process could require much larger energies since Pt^{III} may be a high-energy intermediate of the reduction from Pt^{IV} to Pt^{II}. Since the MMCT transition terminates in an antibonding e_g orbital of Pt^{IV} a large reorganizational energy contributes certainly also to the short wavelength of the MMCT band.

The Pt^{II} to Pt^{IV} OS MMCT absorption of the ion pair $[Pt(NH_3)_5Cl]^{3+}[Pt(CN)_4]^{2-}$ undergoes a further blue shift [50] (Table 1) because the MMCT transition generates now two high-energy Pt^{III} centers. In addition, the oxidation of Pt^{II} is associated with a huge reorganizational energy since new ligands must be attached to the square planar complex.

MMCT of Organometallic Ion Pairs

Organometallic ion pairs which exhibit OS MMCT bands in their electronic spectra have been described only recently. Many salts which contain an oxidizing metal carbonyl or metallocenium cation and a reducing metal carbonyl anion were prepared and characterized a long time ago. However, studies of the optical spectra were not included in the early work. The occurance of OS MMCT absorptions of organometallic ion pairs is not only interesting in its own right but is also of general importance with regard to electron transfer processes in organometallic chemistry [51]. The first example of an organometallic ion pair with an OS MMCT band was reported by Schramm and Zink in 1979. They detected a Co^{-I} to Tl^{I} CT absorption of the ion pair Tl^+ $[Co(CO)_4]^-$ (Table 2) [52].

A very general type of organometallic ion pairs is composed of an oxidizing metal carbonyl cation and a reducing metal carbonyl anion. Some of these

Table 2. Optical OS CT (metal to metal) transitions of organometallic ion pairs

Acceptor	Donor	Abs. λ_{max}/nm	Solvent	Ref.
Tl^+	$[Co(CO)_4]^-$	400	CH_3CN	[52]
$[Co(CO)_3(PPh_3)_2]^+$	$[Co(CO)_4]^-$	386	acetone	[54]
$[Co(C_5H_5)_2]^+$	$[Co(CO)_4]^-$	520	CH_2Cl_2	[59]
$[Co(C_5H_5)_2]^+$	$[Mn(CO)_5]^-$	740	solid	[60]
$[Cr(C_6H_6)_2]^+$	$[Mn(CO)_5]^-$	665	solid	[60]

salts can be prepared by thermal or photochemical disproportionation of dimeric metal carbonyls in the presence of free ligand, e.g. [53]:

$$[Co^0_2(CO)_8] + 2\,PPh_3 \rightarrow [Co^I(CO)_3(PPh_3)_2]^+[Co^{-I}(CO)_4]^- + CO$$

This Co^I/Co^{-I} ion pair is characterized by an OS MMCT absorption in the visible region [54] (Table 2). In addition, a large number of salts which contain the electron acceptors $[M(CO)_6]^+$, $[M(CO)_5L]^+$ and $[M(CO)_4L_2]^+$ with M^I = Mn [55] and Re [56] and L = PPh_3 or L_2 = o-phen and the electron donors $[Co(CO)_4]^-$, $[V(CO)_6]^-$, $[Fe(CO)_3NO]^-$, and $[Mn(CO)_5]^-$ was synthesized. While most of these ions are colorless the salts are generally colored. Although electronic spectra were not recorded the colors originate probably from OS MMCT transitions.

The organometallic ion pairs which are composed of the oxidizing cations $[Co(C_5H_5)_2]^+$ or $[Cr(C_6H_6)_2]^+$ and the reducing anions $[Co(CO)_4]^-$, $[FeH(CO)_4]^-$, $[Cr(C_5H_5)(CO)_3]^-$, $Mn(CO)_5]^-$, and $[V(CO)_6]^-$ are also remarkably colored [57, 58]. In some cases it was confirmed that the colors are due to OS MMCT bands [59, 60] (Table 2).

3.1.2 Ion Pairs Consisting of a Complex and a Non-metallic Counter Ion

Complex to Acceptor CT

There is a large number of oxidizing organic cations which can serve as electron acceptors for reducing metal complex anions. Spectral data on these OS CT transitions are given in Table 3. With regard to the metal complex a distinction between the metal and the ligand as the donor site can be made. Frequently cyano complexes were used as donors [31, 61–66]. A detailed theoretical treatment of the ion pair 1,1′-dimethyl-4,4′-bipyridinium^{2+} $[Fe(CN)_6]^{4-}$ (1,1′-dimethyl-4,4′-bipyridinium^{2+} is also called paraquat^{2+} or methylviologen^{2+}) was published by Curtis, Sullivan, and Meyer [63]. Besides classical cyano complexes also metal carbonyl anions such as $[Co(CO)_4]^-$ [15, 67] and $[M(CO)_6]^-$ with M = V, Nb, and Ta [68–70] are suitable donors. The central metal is the donor site of the cyano and carbonyl complex anions.

Another rather interesting type of reducing complex anion contains the chelating 1,2-ethylenedithiolate ligand (see above). Many ion pairs which are composed of such donating complex anions and organic accepting cations were shown to display OS CT bands [71–74] (Table 3). It seems that for most of these ion pairs the donor site is located at the dithiolate ligands. In the case of the complex $Zn[S_2C_2(CN)_2]_2^{2-}$ [72, 73] this assumption is certainly correct since Zn^{2+} cannot participate in any low-energy CT transitions.

Donor to Complex CT

Many ion pairs which are composed of oxidizing complex cations and non-metallic reducing anions display OS CT bands in their electronic spectra [15]. The acceptor site may be again the metal or the ligand.

Table 3. Optical OS CT (complex to acceptor) transitions of ion pairs consisting of a non-metallic cation and a complex anion

Acceptor	Donor	Abs. λ_{max}/nm	Solvent	Ref.
paraquat^{2+}	$[Fe(CN)_6]^{4-}$	530	H_2O	[63]
Ph_2I^+	$[Fe(CN)_6]^{4-}$	406	CH_3OH	[65]
Ph_2I^+	$[Fe(CN)_5$dimethylsulfoxide$]^{3-}$	333	CH_3OH	[65]
paraquat^{2+}	$[Fe(CN)_5$(imidazole)$]^{3-}$	555	H_2O	[62]
paraquat^{2+}	$[Fe(CN)_5PPh_3]^{3-}$	507	H_2O	[62]
paraquat^{2+}	$[Fe(CN)_5$(pyridine)$]^{3-}$	532	H_2O	[62]
paraquat^{2+}	$[Fe(CN)_5$(dimethylsulfoxide)$]^{3-}$	450	H_2O	[62]
paraquat^{2+}	$[Fe(CN)_5CO]^{3-}$	400	H_2O	[62]
paraquat^{2+}	$[Ru(CN)_6]^{4-}$	416	H_2O	[31]
Ph_2I^+	$[Ru(CN)_6]^{4-}$	340	CH_3OH	[65]
1-ethyl-1-carboxymethylpyridinium$^+$	$[Ru(CN)_6]^{4-}$	442	CH_3OH	[31]
4-methoxyphenyldiazonium$^+$	$[Ru(CN)_6]^{4-}$	375	H_2O	[66]
1-ethyl-2-carboxymethylpyridium$^+$	$[Mn(CN)_5NO]^{3-}$	429	H_2O	[31]
paraquat^{2+}	$[Mn(CN)_5NO]^{3-}$	513	H_2O	[31]
Ph_2I^+	$[Mn(CN)_5NO]^{3-}$	389	CH_3OH	[65]
Ph_2I^+	$[Mo(CN)_8]^{4-}$	370	CH_3OH	[65]
1-ethyl-1-carboxymethylpyriduum$^+$	$[Mo(CN)_8]^{4-}$	444	H_2O	[31]
1-methylchinoxalinium$^+$	$[Mo(CN)_8]^{4-}$	529	H_2O	[31]
paraquat^{2+}	$[Mo(CN)_8]^{4-}$	505	H_2O	[31]
paraquat^{2+}	$[W(CN)_8]^{4-}$	575	H_2O	[31]
1-ethyl-1-carboxymethylpyridium$^+$	$[W(CN)_8]^{4-}$	490	H_2O	[31]
1-methylchinoxylinium$^+$	$[W(CN)_8]^{4-}$	602	H_2O	[31]
Ph_2I^+	$[W(CN)_8]^{4-}$	400	CH_3OH	[65]
pyridinium$^+$	$[Co(CO)_4]^-$	415	butane-2-one	[15, 67]
1-ethyl-4-carbomethoxypyridinium$^+$	$[Ni(mnt)_2]^{2-}$ (mnt = maleonitriledithiolate)	675	solid	[71]
4-cyano-1-methylpyridinium$^+$	$[Ni(mnt)_2]^{2-}$ (mnt = maleonitriledithiolate)	654	solid	[71]
4-cyano-1-ethylpyridinium$^+$	$[Ni(mnt)_2]^{2-}$ (mnt = maleonitriledithiolate)	680	solid	[71]

4-cyano-1-butylpyridinium$^+$	$[Ni(mnt)_2]^{2-}$ (mnt = maleonitriledithiolate)	658	solid	[71]
1-methylpyridinium$^+$	$[Ni(mnt)_2]^{2-}$ (mnt = maleonitriledithiolate)	~634	solid	[71]
1-ethyl-4-carbomethoxypyridinium$^+$	[Ni(1,2-bis(trifluoromethylethylene-1,2-dithiolate)$_2$]$^{2-}$	714	solid	[71]
4-cyano-1-methylpyridinium$^+$	[Ni(1,2-bis(trifluoromethylethylene-1,2-dithiolate)$_2$]$^{2-}$	806	solid	[71]
4-cyano-1-butylpyridinium$^+$	[Ni(1,2-bis(trifluoromethylethylene-1,2-dithiolate)$_2$]$^{2-}$	781	solid	[71]
1-methylpyridinium$^+$	[Ni(1,2-bis(trifluoromethylethylene-1,2-dithiolate)$_2$]$^{2-}$	552	solid	[71]
1-ethyl-4-carbomethoxypyridinium$^+$	$[Co(mnt)_2]^{2-}$	676	solid	[71]
4-cyano-1-butylpyridinium$^+$	$[Co(mnt)_2]^{2-}$	833	solid	[71]
1-methylpyridinium$^+$	$[Co(mnt)_2]^{2-}$	~669	solid	[71]
1-ethyl-4-carbomethoxypyridinium$^+$	$[Cu(mnt)_2]^{2-}$	556	solid	[71]
1-ethyl-4-carbomethoxypyridinium$^+$	$[Zn(mnt)_2]^{2-}$	465	solid	[71]
4-cyano-1-methylpyridinium$^+$	$[Zn(mnt)_2]^{2-}$	588	solid	[71]
4-cyano-1-ethylpyridinium$^+$	$[Zn(mnt)_2]^{2-}$	500	solid	[71]
paraquat^{2+}	$[Zn(mnt)_2]^{2-}$	460	dmso	[72, 73]
paraquat^{2+}	$[Cd(mnt)_2]^{2-}$	460	dmso	[73]
paraquat^{2+}	$[Hg(mnt)_2]^{2-}$	470	dmso	[73]
paraquat^{2+}	[Zn(2,3-chinoxalinedithiolate)$_2$]$^{2-}$	448	dmso	[73]
paraquat^{2+}	[Cd(2,3-chinoxalinedithiolate)$_2$]$^{2-}$	452	dmso	[73]
paraquat^{2+}	[Hg(2,3-chinoxalinedithiolate)$_2$]$^{2-}$	456	dmso	[73]
paraquat^{2+}	$[Zn(ddt)_2]^{2-}$ (ddt = 2-thiooxo-1,3-dithiol-4,5-dithiolate)	655	dmso	[73]
paraquat^{2+}	$[Cd(ddt)_2]^{2-}$	660	dmso	[73]
paraquat^{2+}	$[Hg(ddt)_2]^{2-}$	663	dmso	[73]
1,1′-dioctyl-4,4′-bipyridinium^{2+}	$[Zn(ddt)_2]^{2-}$	660	dmso	[73]
1,1′-dioctyl-4,4′-bipyridinium^{2+}	$[Cd(ddt)_2]^{2-}$	665	dmso	[73]
1,1′-dioctyl-4,4′-bipyridinium^{2+}	$[Hg(ddt)_2]^{2-}$	672	dmso	[73]
paraquat^{2+}	$[Ir(CO)_2(mnt)]^-$	470	CH_3CN	[74]
paraquat^{2+}	$[Ir(P(OPh_3)_3)_2(mnt)]^-$	532	CH_3CN	[74]

Table 4. Optical OS CT (donor to complex) transitions of ion pairs consisting of a complex cation and a non-metallic anion

Acceptor	Donor	Abs. λ_{max}/nm	Solvent	Ref.
$[Co(NH_3)_6]^{3+}$	I^-	272	H_2O	[75, 76]
$[Co(en)_3]^{3+}$	I^-	278	H_2O	[77]
$[Co(en)_3]^{3+}$	SCN^-	285	H_2O	[31]
$[Co(1,2\text{-propanediamine})_3]^{3+}$	I^-	286	H_2O	[77]
$[Co(diethylenetriamine)_2]^{3+}$	I^-	294	H_2O	[77]
$[Co(sep)]^{3+}$	I^-	289	H_2O	[78, 79]
$[Co(sep)]^{3+}$	Br^-	272	H_2O	[78]
$[Co(sep)]^{3+}$	Cl^-	263	H_2O	[78]
$[Co(sep)]^{3+}$	NCS^-	284	H_2O	[78]
$[Co(sep)]^{3+}$	$C_2O_4^{2-}$	275	H_2O	[78, 80]
$[Co(sep)]^{3+}$	BPh_4^-	530	CH_3CN	[81]
$[Ru(NH_3)_5pyridine]^{3+}$	Cl^-	312	H_2O	[83]
$[Ru(NH_3)_5pyridine]^{3+}$	Br^-	338	H_2O	[83]
$[Ru(NH_3)_5pyridine]^{3+}$	SCN^-	400	H_2O	[83]
$[Ru(NH_3)_5pyridine]^{3+}$	I^-	410	H_2O	[83]
$[Ru(NH_3)_5pyridine]^{3+}$	$C_2O_4^{2-}$	409	H_2O	[83
$[Ru(NH_3)_6]^{3+}$	Cl^-	294	H_2O	[82, 83]
$[Ru(NH_3)_6]^{3+}$	Br^-	306	H_2O	[82, 83]
$[Ru(NH_3)_6]^{3+}$	I^-	402	H_2O	[82, 83]
$[Ru(NH_3)_6]^{3+}$	CN^-	406	H_2O	[36]
$[Ru(NH_3)_5(CH_3CN)]^{3+}$	Cl^-	318	H_2O	[83]
$[Ru(NH_3)_5(CH_3CN)]^{3+}$	Br^-	334	H_2O	[83]
$[Ru(en)_3]^{3+}$	I^-	450	H_2O	[84]
		483	solid	[84]
$[Ru(en)_3]^{3+}$	Br^-	370	H_2O	[84]

The majority of observations on OS donor to metal CT absorptions were made with ion pairs which contain cationic amine complexes of Co^{III} [75–81] and Ru^{III} [82–84] as acceptors (Table 4). In addition, fulvalendiyl Co^{III} complexes were used as oxidizing cations [85]. A variety of donor anions such as the halides are suitable. The wavelength of the OS CT bands decreases with decreasing reducing strength of the halide ($I^- > Br^- > Cl^-$). In this context it is certainly of interest that $[Co(NH_3)_6]^{3+}I^-$ was the first ion pair of a metal complex which was reported to show an OS CT absorption [75, 76]. The colors of the salts $[Ir(NH_3)_6]^{3+}$halide$^-$ may be also caused by OS CT bands although the spectra of these ion pairs were not recorded [86].

Instead of an oxidizing metal the coordinated bipy ligand can also act as electron acceptor. For example, OS donor to ligand CT absorptions determine the colors of the salts $[Rh^{III}(bipy)_3]^{3+}X_3^-$ with $X^- = Cl^-$, Br^-, SCN^-, and CN^-. Harris and McKenzie made this observation already in 1963 and suggested a CT transition from X^- to to complex as origin of the colors [87]. Since Rh^{III} is rather redox inert there is little doubt that the OS CT transition terminates indeed in the π^* orbitals of the diimine ligand.

3.2 Neutral Acceptors and/or Donors

Complex to Acceptor CT

The necessary close contact for an OS CT interaction is not only provided by the electrostatic attraction within an ion pair. If the solvent is the donor or acceptor an inimate interaction with a dissolved complex as the acceptor or donor is certainly also guaranteed. While optical OS solvent to complex CT transitions have not yet been identified many reducing complexes are well known to display complex to solvent CT („CTTS = CT to solvent") bands if they are dissolved in oxidizing solvents such as halogenated alkanes or even water [8, 19, 33]. The donor complex can be charged or neutral. Examples are the cations $[Ru^{II}(NH_3)_6]^{2+}$ [88] and $[Co_2(fulvalendiyl)_2]^+$ [85], the anions [89, 90] $[M^{II}(CN)_6]^{4-}$ (M = Fe, Ru) and $[M^{IV}(CN)_8]^{4-}$ (M = Mo, W) and the neutral complexes ferrocene [91, 92] and $[(C_5H_5)\,Fe(CO)]_4$ [93]. The complex to solvent CT bands involving these complexes are usually not well resolved since they appear in the short-wavelength region where they interfere with absorptions of a different origin. Since these complex to solvent CT transitions have been discussed in several reviews [8, 19, 33] any further description is not necessary here.

OS complex to acceptor CT bands were also observed upon association of $[M^0\,(arene)\,(CO)_3]$ (M = Cr, Mo, W) as neutral electron donors and trinitrobenzene (TNB) or tetracyanoethylene (TCNE) as neutral electron acceptors [94–98]. Generally, the CT absorptions appear only for the solid addition compounds while a dissociation into the separate components takes place in solution. The CT interaction occurs by coplanar face to face orientation of TNB with the coordinated arene. The CT energies can be varied systematically by the choice of appropriate substituents at the coordinated arene. Ferrocene and its derivatives were also observed to form addition compounds with TNB and TCNE [96]. These compounds are also characterized by complex to acceptor CT bands in their electronic spectra.

Donor to Complex CT

OS donor to complex CT absorptions which do not involve ion pairs have not yet been identified to our knowledge.

Complex to Complex CT

There is one interesting but rather exotic example which demonstrates the occurance of optical OS CT between neutral complexes. The paramagnetic d^5 complex $V(CO)_6$ is deep green-black but only in the solid state. This color was attributed to an OS MMCT transition from one V^0 to another one [99]. This assumption is supported by the observation that the color disappears upon dissolution. In solution the $V(CO)_5$ molecules are separated from each other and cannot interact electronically. It is of interest that the OS MMCT transition of solid $V^0(CO)_6$ generates the mixed-valence ion pair $[V^I(CO)_6]^+[V^{-I}(CO)_6]^-$ which is not stable but undergoes complete back electron transfer.

4 Photochemistry

An optical OS CT transition is an intermolecular photoredox reaction per definition. However, in most cases the primary redox products undergo a rapid back electron transfer which is favored by the large driving force ΔE of these systems (Fig. 1). In a few cases the primary products were detected and the kinetics of back electron transfer was determined by flash photolysis. The formation of stable photoproducts depends on the competition between back electron transfer and secondary processes. A permanent chemical change takes place if these secondary processes are faster than back electron transfer. This competition can be influenced by a suitable choice or modification of both redox partners. Back electron transfer will be slowed down by increasing its activation energy E_a' (Fig. 1). This can be achieved in two ways. E_a' grows with an increasing reorganizational energy which is associated with a larger horizontal displacement of the potential curve of the primary products. Such large structural changes are encountered when the OS CT transition leads to the population or depopulation of bonding or antibonding instead of non-bonding orbitals. E_a' can also become larger by a decrease of ΔE which is associated with a vertical displacement of the potential curves (Fig. 1). However, at the same time the activation energy E_a for thermal forward electron transfer is lowered. Thermal electron transfer may now occur and interfere with the light-induced process.

A formation of stable photoproducts depends also on the rate of secondary processes which must compete with back electron transfer. For example, Co^{III} amine complexes are well suited as electron acceptors in an irreversible photoreaction since Co^{II} amines undergo a very rapid decay [32]. The dynamics of the solvent cage is also important. A certain fraction of the primary electron transfer products may undergo cage escape before back electron transfer takes place. The primary products which escaped from the cage can react to form stable products.

4.1 Ion Pairs

4.1.1 Complex to Complex Charge Transfer

Photochemical reactions originating from OS LMCT and LLCT excited states are yet unknown. While one example of a reactive OS MLCT state was reported the majority of photoreactions which are induced by OS CT excitation is of the MMCT type.

MLCT

Upon irradiation (λ = 546 nm) of the OS Fe^{II} to bipy MLCT band the cation of the ion pair $[Rh(bipy)_3]^{3+}[Fe(CN)_6]^{4-}$ underwent a photoaquation to $[Rh(bipy)_2(H_2O)_2]^{3+}$ with the quantum yield $\varphi = 2.4 \times 10^{-3}$ [35].

It was suggested that this photoaquation takes place according to the scheme:

$$[Rh^{III}(bipy)_3]^{3+}[Fe^{II}(CN)_6]^{4-} \xrightarrow{hv}$$

$$[Rh^{III}(bipy)_2(bipy^-)]^{2+}[Fe^{III}(CN)_6]^{3-}$$

$$[Rh(bipy)_3]^{2+}[Fe(CN)_6]^{3-} \rightarrow [Rh(bipy)_3]^{2+} + [Fe(CN)_6]^{3-}$$

$$[Rh(bipy)_3]^{2+} \rightarrow [Rh(bipy)_2]^{2+} + bpy$$

$$[Rh(bipy)_2]^{2+} + [Fe(CN)_6]^{3-} + 2\,H_2O \rightarrow$$

$$[Rh(bipy)_2(H_2O_2]^{3+} + [Fe(CN)_6]^{4-}$$

The ion pair generated by outer-sphere MLCT excitation may diffuse apart. The third equation describes only the fact that $[Rh(bpy)_3]^{2+}$ is known to release a bpy ligand. The mechanism of this reaction is not quite clear. Various possibilities, including a disproportionation, are feasible. Finally, electron transfer and subsequent formation of $[Rh(bipy)_2(H_2O)_2]^{3-}$ is the last step of this mechanism. The low quantum yield of the overall reaction is probably determined by the rapid thermal reversal of the optical CT transition. This back electron transfer competes with the diffusion step. The ion pairs $[Rh(bipy)_3]^{3+}[M(CN)_6]^{4-}$ with M = Ru, Os seem to be also light-sensitive upon outer-sphere MLCT excitation. However, in this case the CT bands occur at shorter wavelength and thus are overlapping with intramolecular absorption bands of the complexes. For this reason a selective outer-sphere MLCT excitation could not be achieved.

MMCT

The ion pair $[Co(NH_3)_6]^{3+}[Ru(CN)_6]^{4-}$ in dimethylsulfoxide underwent a photoredox reaction upon Ru^{II} to Co^{III} MMCT excitation ($\varnothing = 0.034$ at $\lambda_{irr} = 366$ nm) [41]:

$$[Co^{III}(NH_3)_6]^{3+}[Ru^{II}(CN)_6]^{4-} \xrightarrow{hv}$$

$$[Co^{II}(NH_3)_6]^{2+}[Ru^{III}(CN)_6]^{3-}$$

$$[Co^{II}(NH_3)_6]^{2+} \rightarrow Co^{2+} + 6\,NH_3$$

Electron transfer to $[Co(NH_3)_6]^{3+}$ requires a large reorganizational energy due to the population of an antibonding e_g orbital. As a result the activation energy for back electron transfer E_a' should be also rather large. Moreover, the decomposition of $[Co^{II}(NH_3)_6]^{2+}$ is a rapid process which occurs with a rate constant larger than $10^6\ s^{-1}$ [32]. The low quantum yield of Co^{2+} formation shows, however, that the competition by back electron transfer is still very efficient.

When the ammonia ligands of $[Co(NH_3]_6]^{3+}$ are replaced by ethylenediamine (en) and its derivatives (1,2-diaminopropane and 1,2-cyclohexanediamine) the

Ru^{II} to Co^{III} MMCT excitation of the aqueous ion pairs leads to a different result [12, 42]:

$$[Co^{III}(en)_3]^{3+}[Ru^{II}(CN)_6]^{4-} \xrightarrow{h\nu} [Co^{II}(en)_3]^{2+}[Ru^{III}(CN)_6]^{3-}$$

$$[Co^{II}(en)_3]^{2+}[Ru^{III}(CN)_6]^{3-} \xrightarrow[-en]{+H_2O}$$

$$[(en)_2(H_2O)Co^{II}-NC-Ru^{III}(CN)_5]^-$$

$$[(en)_2(H_2O)Co^{II}-NC-Ru^{III}(CN)_5]^- \rightarrow$$

$$[(en)_2(H_2O)Co^{III}-NC-Ru^{II}(CN)_5]^-$$

The primary reduction product $[Co(en)_3]^{2+}$ is also substitutionally labile but does not decompose as fast as $[Co(NH_3)_6]^{2+}$ [100, 101]. $[Co(en)_3]^{2+}$ is then substituted by $[Ru(CN)_6]^{3-}$. Finally, the thermal electron transfer from Co^{II} to Ru^{III} by an IS process generates the stable cyanide-bridged complex. The binuclear complex was not isolated but its formation was suggested on the basis of spectral data. The course of this photoreaction is not surprising. It is well known that $[Co^{II}(CN)_5]^{3-}$ reacts thermally with $[M^{II}(CN)_6]^{4-}$ (M = Fe, Ru, Os) by an IS electron transfer to yield the binuclear complexes $[(NC)_5Co^{III}-NC-M^{II}(CN)_5]^{6-}$ [102–104).

The aqueous pair $[Co(sep)]^{3+}[Ru(CN)_6]^{4-}$ with sep = sepulchrate does not undergo any permanent chemical change upon MMCT excitation since the primary reduction product $[Co(sep)]^{2+}$ is kinetically stable due to the nature of the cage-type ligand sep [12, 42]. Back electron transfer is now the only secondary process.

When Ru^{II} is replaced by Fe^{II} ΔE of the ion pair $[Co(en)_3]^{3+}[Fe(CN)_6]^{4-}$ becomes smaller by 0.5 V. Consequently, the energy of the MMCT transition (Table 1) as well as the activation energy for thermal electron transfer E_a decreases. It has been shown indeed that the formation of a binuclear cyanide-bridged complex, most likely $[(en)_2(H_2O)Co^{III}-NC-Fe^{II}(CN)_5]^-$, does not only occur as a photochemical but also as a slow thermal reaction [105]. In the presence of excess chloride the photolysis leads to the formation of $[Cl(en)_2Co^{III}NC-Fe^{II}(CN)_5]^{2-}$ as a stable product [106, 107]. The incorporation of chloride into the complex takes certainly place prior to back electron transfer. In this experiment the observation of an OS MMCT absorption was not reported. It was suggested that the reaction could originate from a OS MMCT state which might have been populated from a ligand field excited state of $[Co(en)_3]^{3+}$.

Octahedral ammine complexes of the d^5 metals Ru^{III} and Os^{III} can also serve as electron acceptors. However, in distinction to the d^6 metal Co^{III} the reduction of Ru^{III} and Os^{III} by CT excitation does not require a large reorganizational energy since the acceptor orbitals (t_{2g}) are non-bonding. Their population has thus not a large effect on the metal-ligand distance. The activation energy for back electron transfer is presumably rather small. In addition, ammine complexes of Ru^{II} and Os^{II} as primary reduction products are less labile than the ammine complexes of Co^{II}. For these reasons the ion pairs $[M^{III}(NH_3)_5L]^{n+}[M^{II}(CN)_5]^{4-}$ with M^{III} = Ru^{III}, Os^{III} and M^{II} = Fe^{II}, Ru^{II}, Os^{II} are not expected to undergo an efficient formation of stable products upon OS MMCT excitation. Generally, this expectation has been confirmed but in some cases a photoactivity was observed.

The aqueous homonuclear ion pair $[Ru^{III}(NH_3)_5Cl]^{2+}[Ru^{II}(CN)_6]^{4-}$ underwent a photolysis upon MMCT excitation at $\lambda = 546$ nm with $\varphi = 0.002$ [47]. The reaction proceeds according to the following scheme:

$$[Ru^{III}(NH_3)_5Cl]^{2+}[Ru^{II}(CN)_6]^{4-} \xrightarrow{h\nu} [Ru^{II}(NH_3)_5Cl]^{+}[Ru^{III}(CN)_6]^{3-}$$

$$[Ru^{II}(NH_3)_5Cl]^{+}[Ru^{III}(CN)_6]^{3-} \rightarrow$$
$$[Ru^{II}(NH_3)_5Cl]^{+} + [Ru^{III}(CN)_6]^{3-}$$

$$[Ru^{II}(NH_3)_5Cl]^{+} + H_2O \rightarrow [Ru^{II}(NH_3)_5(H_2O)]^{2+} + Cl^{-}$$

$$[Ru^{II}(NH_3)_5H_2O]^{2+} + [Ru^{II}(CN)_6]^{4-} \rightarrow$$
$$[(NH_3)_5Ru^{II}{-}NC{-}Ru^{II}(CN)_5]^{2-}$$

$$[(NH_3)_5Ru^{II}{-}NC{-}Ru^{II}(CN)_5]^{2-} + [Ru^{III}(CN)_6]^{3-} \rightarrow$$
$$[(NH_3)_5Ru^{III}{-}NC{-}Ru^{II}(CN)_5]^{-} + [Ru^{II}(CN)_6]^{4-}$$

The ion pair, generated by MMCT excitation may diffuse apart. $[Ru^{II}(NH_3)_5Cl]^{+}$ aquates rapidly with $k = 5\ s^{-1}$. Substitutionally labile $[Ru^{II}(NH_3)_5H_2O]^{2+}$ reacts with $[Ru(CN)_6]^{4-}$ which is present in large excess. The formation of the binuclear complex is certainly facilitated by the high opposite charges of the reacting ions. Finally, electron transfer restores ruthenium to its stable oxidation states. The low quantum yield of the overall reaction is most likely determined by the extremely rapid thermal reversal of the MMCT transition, which competes with the diffusion apart from the primary electron transfer products. Product formation occurs also thermally indicating a small activation energy (~22 kcal/mol) for thermal electron transfer.

The aqueous ion pairs $[Os^{III}(NH_3)_5Cl]^{2+}[M^{II}(CN)_6]^{4-}$ with M = Fe, Ru and Os undergo photoreactions which are quite analogous to that of $[Ru^{III}(NH_3)_5Cl]^{2+}$-$[Ru^{II}(CN)_6]^{4-}$. Upon M^{II} to Os^{III} MMCT excitation ($\lambda_{irr} = 405$ nm) the binuclear complexes $[(NH_3)_5Os^{III}{-}NC{-}M^{II}(CN)_5]^{-}$ are formed [12, 42]. The quantum yields ($\varphi = 0.12$ for M = Fe, 0.04 for Ru and Os) are larger than that of the Ru^{II}/Ru^{III} ion pair.

OS Fe^{II} to Ru^{III} MMCT excitation ($\lambda_{irr} = 1060$ nm) of the aqueous ion pair $[Ru^{III}(NH_3)_5(py)]^{3+}[Fe^{II}(CN)_6]^{4-}$ with py = pyridine does not yield stable products. However, flash photolysis revealed some interesting details of the reversible photoreaction [40, 108]. The Franck-Condon excited MMCT state $[Ru^{II}(NH_3)_5$-$py]^{2+}[Fe^{III}(CN)_6]^{3-}$ does not undergo a vibrational relaxation with unit efficiency but returns partially to the Ru^{III}/Fe^{II} ground state. The equilibrated Ru^{II}/Fe^{III} MMCT state undergoes also a back electron transfer and a competing cage escape. Finally, the separated ions $[Ru^{II}(NH_3)_5py]^{2+}$ and $[Fe^{III}(CN)_6]^{3-}$ regenerate the ion pair. Back electron transfer in this ion pair restores the stable oxidation states.

Organometallic Ion Pairs

Many salts consisting of an oxidizing metal carbonyl cation and a reducing metal carbonyl anion were prepared and characterized. They are expected to display long-wavelength OS MMCT absorptions. This was confirmed for the ion pair $[Co(CO)_3(PPh_3)_2]^+[Co(CO)_4]^-$ [54] (Table 2). The photochemistry of this ion pair is rather complicated upon short-wavelength irradiation which leads to the excitation of the cation [109]. When the light is absorbed by the MMCT band at longer wavelength (λ_{irr} = 405 nm) a radical pair is formed. The processes which are initiated by the MMCT transition can be explained by the following mechanism [54]:

$$[Co^{+I}(CO)_3(PPh_3)_2]^+[Co^{-I}(CO)_4]^- \xrightarrow{h\nu}$$

$$[Co^0(CO)_3(PPh_3)_2][Co^0(CO)_4]$$

$$[Co^{+I}(CO)_3(PPh_3)_2][Co^0(CO)_4] \rightarrow [Co^0(CO)_3(PPh_3)_2] + [Co^0(CO)_4]$$

$$[Co^0(CO)_3(PPh_3)_2] \rightarrow [Co^0(CO)_3(PPh_3)] + PPh_3$$

$$[Co^0(CO)_4 + PPh_3 \rightarrow [Co^0(CO)_3PPh_3] + CO$$

$$2\,[Co^0(CO)_3(PPh_3) \rightarrow [(PPh_3)(CO)_3Co{-}Co(CO)_3(PPh_3)]$$

The primary radicals may diffuse apart. These radicals are certainly labile toward substitution or dissociation. Finally, the radical $[Co(CO)_3(PPh_3)]$ is formed which dimerizes to the product. The low quantum yield of the photoreaction (φ = 0.012) may be due to a competing reversal of some of these processes including back electron transfer within the primary radical pair. It is quite interesting that product formation occurs also thermally at elevated temperatures [110]. This indicates a relatively low activation energy E_a (Fig. 1). In this context it is rather important to pay attention to the general significance of this thermal reaction in organometallic chemistry [51]. With regard to the photochemistry we would like to emphasize the relationship between the light-induced formation and homolytic cleavage of metal-metal bonds:

$$[M^{+I}(CO)_nL]^+[M^{-I}(CO)_n] \xrightleftharpoons{MMCT} 2[\cdot M^0(CO)_n] + L \xrightleftharpoons{\sigma\sigma^*}$$

$$[(OC)_nM^0{-}M^0(CO)_n] + L$$

Both photoreactions, electron transfer by MMCT excitation and metal-metal bond splitting by $\sigma\sigma^*$ excitation [19, 111, 112], generate the same or similar radicals.

Instead of metal carbonyl cations, oxidizing metallocenium cations can be also used as electron acceptors. Upon MMCT excitation (λ_{irr} > 520 nm) of $[Co^{III}(C_5H_5)_2{}^+[Co^{-I}(CO)_4]^-$ in THF, cobaltocene and $[Co(CO)_4]$ are formed as a radical pair which undergoes a rapid back electron transfer to the starting ion pair [59]. In the presence of phosphites the formation of stable products takes place.

$$[Co^{III}(C_5H_5)_2]^+[Co^{-I}(CO)_4]^- \xrightarrow{h\nu} [Co^{II}(C_5H_5)_2][Co^0(CO)_4]$$

$$[Co^{II}(C_5H_5)_2]^+ [Co^0(CO)_4] \rightarrow [Co^{II}(C_5H_5)] + [Co^0(CO)_4]$$

$$[Co^0(CO)_4] + P(OR)_3 \rightarrow [Co^0(CO)_3P(OR)_3] + CO$$

$$[Co^{II}(C_5H_5)_2] + [Co^0(CO)_3P(OR)_3 \rightarrow$$

$$[Co^{III}(C_5H_5)_2]^+ [Co^{-I}(CO)_3P(OR)_3]^-$$

The $[Co(CO)_4]$ radicals which escape the primary radical pair are substituted before back electron transfer occurs. In the presence of phosphines the reaction takes a different course:

$$[Co^{III}(C_5H_5)_2]^+[Co^{-I}(CO)_4]^- + PR_3 \xrightarrow{h\nu}$$

$$[Co^{II}(C_5H_5)_2] + 1/2[Co_2^0(CO)_6(PR_3)_2]$$

The dimerization of the substituted radical $[Co(CO)_3PR_3]$ is now apparently faster than back electron transfer.

Upon MMCT excitation ($\lambda_{irr} > 580$ nm) the ion pair $[Co^{II}(C_5H_5)_2]^+ [Mn^{-I}$-$(CO)_5]^-$ undergoes an analogous photoreaction in the absence of an entering ligand [60]:

$$[Co^{III}(C_5H_5)_2]^+[Co^{-I}(CO)_4]^- \xrightarrow{h\nu} [Co^{II}(C_5H_5)_2] + 1/2[Mn_2^0(CO)_{10}]$$

It is quite surprising that this photoreaction proceeds in a KBr matrix since the mobility and hence the cage escape of $[Mn(CO)_5]$ radicals should be hindered in this medium.

The ion pair $[Co(C_5H_5)_2]^+[Mn(CO)_5]^-$ is thermally rather stable in distinction to $[Cr(C_6H_6)_2]^+ [Mn(CO)_5]^-$ which undergoes the electron transfer under ambient conditions [58, 60]:

$$[Cr^I(C_6H_6)_2]^+[Mn^{-I}(CO)_5]^- \rightarrow [Cr^0(C_6H_6)_2] + 1/2[Mn_2^0(CO)_{10}]$$

This different behavior is quite unexpected since the potentials of the redox couples $[Co(C_5H_5)_2]^{+/0}$ and $[Cr(C_6H_6)_2]^{+/0}$ and consequently ΔE of both ion pairs are very similar. However, the cation $[Cr^I(C_6H_6)_2]^+$ which contains a d^5 metal accepts the electron in a non-bonding a_{1g} orbital. This is associated with a rather small reorganizational energy and therefore also low activation energy for thermal electron transfer. On the contrary, $[Co(C_5H_5)_2]^+$ with a d^6 metal must accept the electron into an antibonding e_{1g} orbital which requires a much larger reorganizational and activation energy.

4.1.2 Ion Pairs Consisting of a Complex and a Non-metallic Counter Ion

Complex to Acceptor CT

The majority of ion pairs which are composed of an organic cation as acceptor and a metal complex anion as donor have not been observed to be light sensitive upon OS CT exciation. Secondary reactions of the primary redox products are apparently too slow to compete with back electron transfer which restores the starting ion pair. Kinetic data were obtained for the ion pair 1,1′-dimethyl-4,4′-bipyridinium^{2+}[Zn(maleonitriledithiolate)$_2$]$^{2-}$ [72]. OS CT excitation of the ion pair leads to an electron transfer from the coordinated ligand to the cation. The primary photoproduct 1,1′-dimethyl-4,4′-bipyridinium^{+}[Zn(maleonitriledithiolate)$_2$]$^{-}$ was detected by flash photolysis. It regenerates the starting ion pair by back electron transfer with the second order rate constant $k = 3.6 \times 10^9\ M^{-1}\ s^{-1}$.

When Ph_2I^+ and RN_2^+ are used as acceptors an OS CT excitation results in a permanent chemical change since these cations undergo an irreversible reduction. The decay of the radicals $Ph_2I\cdot$ and $R{-}N_2\cdot$ is apparently a very rapid process which competes successfully with their reoxidation.

Hennig and his group investigated the photochemistry of nonaqueous solutions of ion pairs which consist of the diphenyliodonium cation as acceptor and the cyano complexes $[Fe^{II}(CN)_5DMSO]^{3-}$, $[Ru^{II}(CN)_6]^{4-}$, $[Mo^{IV}(CN)_8]^{4-}$, $[W^{IV}(CN)_8]^{4-}$, and $[Mn^{I}(CN)_5NO]^{3-}$ as donors [64, 65]. As an example, upon OS complex to acceptor CT excitation ($\lambda_{irr} = 475$) the ion pair $Ph_2I^+[Mo(CN)_8]^{4-}$ reacts according to the following equations:

$$Ph_2I^+[Mo^{IV}(CN)_8]^{4-} \xrightarrow{h\nu} Ph_2I\cdot + [Mo^{V}(CN)_8]^{3-}$$

$$Ph_2I\cdot \rightarrow Ph{-}I + Ph\cdot$$

The diphenyliodonium radical decomposes to iodobenzene and a phenyl radical which undergoes further reactions. The other ion pairs undergo the same type of reaction. The quantum yields are rather high (0.3 to 0.8)]

Arenediazonium cations are reduced thermally by $[Fe(CN)_6]^{4-}$ in aqueous solution at room temperature [113]. The irreversible reduction yields N_2 and phenyl radicals. The slowest electron transfer was observed with the *p*-methoxybenzenediazonium cation ($k = 0.71\ M^{-1}\ s^{-1}$). Unfortunately, due to their thermal instability the ion pairs $R{-}C_6H_4{-}N_2^+[Fe(CN)_6]^{4-}$ are not well suited to study their electronic spectra and photoreactions. However, much more stable ion pairs should be formed if $[Fe(CN)_6]^{4-}$ is replaced by the less reducing anion $[Ru(CN)_6]^{4-}$. The ion pair *p*-$CH_3O{-}C_6H_4{-}N_2^+[Ru(CN)_6]^{4-}$ is indeed rather stable but undergoes an irreversible photoreaction upon OS CT excitation ($\lambda_{irr} = 405$ nm) [66]. $[Ru(CN)_6]^{3-}$ was formed with the quantum yield $\varphi = 0.02$.

It is suggested that the photolysis proceed according to the equation:

$$CH_3O{-}C_6H_4{-}N_2^+[Ru^{II}(CN)_6]^{4-} \xrightarrow{h\nu}$$
$$CH_3O{-}C_6H_4N_2[Ru^{III}(CN)_6]^{3-}$$
$$CH_3O{-}C_6H_4N_2 \rightarrow CH_3O{-}C_6H_4\cdot + N_2$$

The fate of the *p*-methoxyphenyl radical was not investigated.

Donor to Complex CT

A number of studies on the photochemistry of ion pairs of the type $[M^{III}(NH_3)_5L]^{n+}$ X^- with M = Co, Ru and X = halide and other anions has been carried out. Ford and his group investigated the photolysis of Ru^{III} ammines following X^- to Ru^{III} OS CT excitation [83]. These ion pairs are not expected to be particularly light-sensitive upon OS CT excitation since the primary reduction product $[Ru^{II}(NH_3)_5L]^{+(n-1)}$ is fairly stable. Back electron transfer may be then much faster than any other secondary reaction. Generally, this expectation was confirmed. The flash photolysis of $[Ru^{III}(NH_3)_5pyridine]^{3+}Cl^-$ did not even yield any transient indicating an extremely rapid regeneration of the starting ion pair. When X^- was Br^- or I^- the formation of Ru^{II} intermediates was observed. However, back electron transfer was still very fast. With X = I a low-yield photosubstitution took place upon continuous irradiation of the OS CT band. In the case of the ion pair $[Ru^{III}(NH_3)_5]^{3+}C_2O_4^{2-}$ an efficient irreversible photoreduction to $[Ru^{II}(NH_3)_5pyridine]^{2+}$ took place (φ = 0.35 at λ_{irr} = 405 nm). This is certainly due to the rapid decay of the oxidized oxalate.

When Co^{III} ammines are used as acceptors OS CT excitation is generally associated with a permanent reduction to Co^{III} (see 4.1.1). Historically, the first study on a photoreaction of a metal complex induced by OS CT excitation was carried out by Adamson and Sporer in 1958 [114]. They found that light absorption by an OS (I^- to Co^{III}) CT bad of the aqueous ion pair $[Co(NH_3)_6]^{3+}I^-$ led to a redox decomposition (φ = 0.77 at λ_{irr} = 370 nm). The photolysis may proceed according the following equation:

$$[Co^{III}(NH_3)_6]^{3+}I^- \xrightarrow{h\nu} [Co^{II}(NH_3)_6]^{2+}I \rightarrow Co^{2+} + 6\,NH_3 + 1/2I_2$$

An analogous photoreaction of $[Co(NH_3)_6]^{3+}(BPh_4)^-$ was investigated by Hannig and his group [115, 116]. In this case the formation of stable redox products upon OS CT excitation is not only favored by the facile decay of $[Co(NH_3)_6]^{2+}$ but also by an efficient irreversible decomposition of the BPh_4 radicals which split off phenyl radicals. Essentially the same behavior is shown by the ion pairs $[Co^{III}(NH_3)_5Y]^{2+}X^-$ with Y^- = CH_3COO^-, NO_2^-, Br^-, Cl^-, F^-, N_3^-, and NO_3^- and X^- = BPh_4^- and I^-.

When $[Co(NH_3)_6]^{3+}$ is replaced by $[Co(en)_3]^{3+}$ an OS X^- to Co^{III} CT excitation may not be expected to lead to an efficient production of Co^{III} since $[Co^{II}(en)_3]^{2+}$ decays much slower than $[Co^{II}(NH_3)_6]^{2+}$ [100, 101] (see above). However, if X^-

undergoes an oxidative decomposition the production of Co^{2+} can still take place with high quantum yields. For example, Co^{2+} is formed with $\varphi = 0.13$ ($\lambda_{irr} = 313$ nm) upon OS CT excitation of $[Co(en)_3]^{3+}HC_2O_4^-$ [80]. The ion pair $[Co(en)_3]^{3+}htc^-$ with htc^- = bis(2-hydroxyethyl)dithiocarbarmate undergoes the same type of photoredox reaction [117]. In the presence of an excess of htc^- Co^{2+} forms a stable complex with this ligand. This Co^{II}htc complex is now able to reduce $[Co(en)_3]^{3+}$ thermally.

Consequently a chain reaction occurs and the observed quantum yield for Co^{2+} production can exceed unity. Analogous photoreactions were found for the ion pairs of $[Co(NH_3)_6]^{3+}$, $[Co(NH_3)_4(1,2\text{-propanediamine})]^{3+}$, [Co(1,2-cyclohexanediamine)$_3$]$^{3+}$ and [Co(diethylenetriamine)$_2$]$^{3+}$ with htc^- as counter ion.

The acceptor cation $[Co^{III}(sep)]^{3+}$ does not undergo an irreversible reduction since the Co^{II} complex is kinetically stable due to the cage-type nature of the sepulchrate ligand. It is then not surprising that in deoxygenated, neutral solutions OS CT excitation ($\lambda_{irr} = 313$ nm) of $[Co(sep)]^{3+}I^-$ does not lead to any permanent chemical change. The primary redox product $[Co^{II}(sep)]^{2+}$ and iodine simply regenerate the starting ion pair [78, 79]. However, in acidic medium $[Co(sep)]^{2+}$ is not any more stable. The decomposition competes now with its reoxidation by iodine. As a result Co^{2+} and I_2 are produced although with small quantum yields. In the presence of air $[Co(sep)]^{2+}$ can be intercepted by O_2. In this case $[Co(sep)]^{3+}$ acts as a sensitizer for the photoassisted oxidation of iodide by O_2:

$$4I^- + O_2 + 4H^+ \rightarrow 2I_2 + 2\,H_2O$$

If deoxygenated solutions of $[Co(sep)]^{3+}X^-$ with $X^- = BPh_4^-$ [81] or oxalate [78, 80] are irradiated into the OS CT band the formation of $[Co^{II}(sep)]^{2+}$ is observed since the anions are oxidized irreversibly. The photoreduction of aqueous $[Co(sep)]^{3+}$ by oxalate can be used for the generation of H_2 since in the presence of colloidal platinum $[Co(sep)]^{2+}$ is able to reduce water. $[Co(sep)]^{3+}$ sensitizes thus the reaction [78, 80]:

$$H_2C_2O_4 \rightarrow 2\,CO_2 + H_2$$

The organometallic ion pair $[Co_2(C_{10}H_8)_2]^{2+}BPh_4^-$ represents a further interesting example of photoreactivity following OS donor to complex CT excitation [85]. The photolysis of the ion pair in CH_3CN is associated with the reduction of the complex to $[Co_2(C_{10}H_8)_2]^+$. Since this cation is stable product formation occurs only by the irreversible decay of the BPh_4 radical.

4.2 Neutral Acceptors and/or Donors

As discussed in Sect. 3.2 the electronic spectra of reducing complexes dissolved in oxidizing solvents display frequently OS complex to solvent CT bands. Light absorption by such CT bands is associated with the oxidation of the complex. The solvent accepts the electrons either by an irreversible reduction (e.g. halogenated alkanes) or by the formation of solvated electrons (e.g. H_2O) which can be intercepted by suitable scavengers such as N_2O. The complexes are then also irreversibly photooxidized. Anionic cyanide complexes such as $[Fe(CN)_6]^{4-}$, $[Ru(CN)_6]^{4-}$ or $[Mo(CN)_8]^{4-}$ [89, 90] and neutral organometallic complexes such as ferrocene [91, 92] have been used as electron donors. But also reducing cations such as $[Ru(NH_3)_6]^{2+}$ [118] and $[Co_2(C_{10}H_8)_2]^+$ [85] are photooxidized upon OS complex to solvent CT excitation. Since this subject has been covered by several reviews [8, 19, 33] it is not further discussed here.

To our knowledge there is not any other report on the photochemistry initiated by OS CT excitation not involving ion pairs or the solvent. Surely, such systems will be discovered in the future. For example, the green addition complexes formed between chloranil or tetracyanoethylene as acceptors and tricarbonyltoluenechromium as donor were reported to be photochemically unstable [94]. However, the nature of this light-sensitivity was not explored.

5 Outlook and Conclusion

Light-induced reactions which originate from OS CT excited states have been shown to play an important role in the field of photochemistry of coordination compounds. Since the majority of observations on this subject was reported only recently it is expected that many more examples of reactive OS CT states will be discovered in the near future. (After completion of this review an extensive publication on the spectroscopy and photochemistry of organometallic ions pairs was published by Bockman and Kochi [119]). We can anticipate the design of new photoactive OS CT systems for applications in industrial photochemistry. The observations on reactive OS CT states involving metal complexes are not only interesting in their own right, they are also an important supplement to the research on excited state electron transfer. Since optical OS CT is intimately related to thermal electron transfer our increasing knowledge on this subject will also contribute to a better understanding of the mechanism of thermal redox reactions.

6 References

1. Balzani V, Bolletta F, Gandolfi MT, Maestri M (1978) Top. Curr. Chem. 75: 1
2. Meyer TJ (1978) Acc. Chem. Res. 11: 94
3. Sutin N, Creutz C (1980) Pure Appl. Chem. 52: 2717
4. Meyer TJ (1983) Progr. Inorg. Chem. 30: 389
5. Sutin N, Creutz C (1983) J. Chem. Ed. 60: 809
6. Serpone N (1988) In: Fox MA, Chanon M (eds) Photoinduced electron transfer. Elsevier, Amsterdam, part D, p 47
7. Balzani V, Scandola F (1988) In: Fox MA, Chanon M (eds) Photoinduced electron transfer. Elsevier, Amsterdam, part D, p 148
8. Gianotti C, Gaspard S, Kransz P (1988) In: Fox MA, Chanon M (eds) Photoinduced electron transfer. Elsevier, Amsterdam, part D, p 200

9. Grätzel M (1988) In: Fox MA, Chanon M (eds) Photoinduced electron transfer. Elsevier, Amsterdam, part D, p 394
10. Meyer TJ (1989) Acc. Chem. Res. 22: 163
11. Grätzel M (1989) Heterogeneous photochemical electron transfer. CRC Press, Boca Raton
12. Vogler A, Osman AH, Kunkely H (1985) Coord. Chem. Rev. 64: 159
13. Hennig H, Rehorek D, Archer RD (1985) Coord. Chem. Rev. 61: 1
14. Balzani V, Sabbatini N, Scandola F (1986) Chem. Rev. 86: 319
15. Hennig H, Rehorek D, Billing R (1988) Comments Inorg. Chem. 8: 163
16. Stein CA, Taube H (1978) J. Am. Chem. Soc. 100: 1635
17. Vogler A (1988) In: Fox MA, Chanon M (eds) Photoinduced electron transfer. Elsevier, Amsterdam, part D, p 179
18. Lever ABP (1984) Inorganic electronic spectroscopy, Elsevier, Amsterdam
19. Geoffroy GL, Wrighton MS (1979) Organometallic photochemistry, Academic, New York
20. Taube H (1978) Ann. N.Y. Acad. Sci. 313: 483
21. Meyer TJ (1978) Ann. N.Y. Acad. Sci. 313: 496
22. Brown D (ed) (1980) Mixed-valence compounds, Reidel, Dordrecht
23. Creutz C (1980) Progr. Inorg. Chem. 30: 1
24. Hush NS (1967) Progr. Inorg. Chem. 8: 391
25. Vogler A, Kunkely H (1989) Z. Naturforsch. 44b: 132
26. Vogler A, Kunkely H (1990) Comments Inorg. Chem. 9: 201
27. Hush NS (1961) Trans. Faraday Soc. 57: 557
28. Nush NS (1968) Electrochim. Acta 13: 1005
29. Powers MJ, Meyer TJ (1980) J. Am. Chem. Soc. 102: 1289
30. Hennig H, Billing R, Benedix R (1986) Monatshefte Chem. 117: 51
31. Hennig H, Benedix R, Billing R (1986) J. Prakt. Chem. 328: 829
32. Simic M, Lilie J (1974) J. Am. Chem. Soc. 96: 291
33. Fox M (1975) In: Adamson AW, Fleischauer PD (eds) Concepts of inorganic chemistry, Wiley, New York, p 333.
34. Jørgensen CK (1963) Acta Chem. Scand. 17: 1034
35. Vogler A, Kunkely H (1987) Inorg. Chem. 26: 1819
36. Vogler A, Kunkely H (1988) Inorg. Chim. Acta 144: 149
37. Vogler A, Kunkely H (1986) J. Chem. Soc. Chem. Commun.: 1616
38. Hennig H, Rehorek A, Ackermann M, Rehorek D, Thomas P (1983) Z. Anorg. Allg. Chem. 496: 186
39. Hennig H, Rehorek A, Rehorek D, Thomas P (1984) Inorg. Chim. Acta 86: 41
40. Haim A (1985) Comments Inorg. Chem. 4: 113
41. Vogler A, Kisslinger J (1982) Angew. Chem. Int. Ed. Engl. 21: 77
42. Osman AH (1987) Inter- und intramolekulare Photoredoxreaktionen von Übergangsmetallkomplexen. Thesis, Universität Regensburg, Regensburg
43. Vogler A, Kunkely H (1975) Ber. Bunsenges. Phys. Chem. 79: 83
44. Curtis JC, Meyer TJ (1978) J. Am. Chem. Soc. 100: 6284
45. Toma HE (1980) J. Chem. Soc. Dalton: 471
46. Curtis JC, Meyer TJ (1982) Inorg. Chem. 21: 1562
47. Vogler A, Kisslinger J (1982) J. Am. Chem. Soc. 104: 2311
48. Sabbatini N, Bonazzi A, Ciano M, Balzani V (1984) J. Am. Chem. Soc. 106: 4055
49. Sabbatini N, Balzani V (1985) J] Less-Common Met. 112: 381
50. Vogler A, Kunkely H (1988) Inorg. Chim. Acta 150: 3
51. Lee KY, Kochi JK (1989) Inorg. Chem. 28: 567
52. Schramm C, Zink JI (1979) J. Am. Chem. Soc. 101: 4554
53. Hieber W, Vohler O, Braun G (1958) Z. Naturforsch. 13b: 192
54. Vogler A, Kunkely H (1988) Organometallics 7: 1449
55. Kruck T, Höfler M (1964) Chem. Ber. 97: 2289
56. Kruck T, Höfler M, Noack M (1966) Chem. Ber. 99: 1153
57. Fischer EO, Kögler HP (1956) Angew. Chem. 68: 462
58. Hieber W, Schropp W (1960) Chem. Ber. 93: 455
59. Bockman TM, Kochi JK (1988) J. Am. Chem. Soc. 110: 1294

60. Kunkely H, Vogler A (1989) J. Organomet. Chem. 372: C29
61. Nakahara A, Wang JH (1963) J. Phys. Chem. 67: 496
62. Toma HE (1979) Can. J. Chem. 57: 2079
63. Curtis JC, Sullivan BP, Meyer TJ (1980) Inorg. Chem. 19: 3833
64. Rehorek D, Hantschmann A, Salvetter J, Hennig H (1979) J. Prakt. Chem. 328: 159
65. Billing R, Rehorek D, Salvetter J, Hennig H (1988) Z. Anorg. Allg. Chem. 557: 234
66. Kisslinger J, Vogler A (unpublished results)
67. Borovnikov MS, Geosdovskij GN, Rybakov VA, Tarasov BP (1982) Zhurn. Obskh. Khim. 52: 331
68. Calderazzo F, Pampaloni G, Lanfranchi M, Pelizzi G (1985) J. Organomet. Chem. 296: 1
69. Calderazzo F, Pampaloni G (1986) J. Organomet. Chem. 303: 111
70. Calderazzo F, Pampaloni G, Pelizzi G, Vitali F (1988) Organometallics 7: 1083.
71. Dance IG, Solstad PJ (1973) J. Am. Chem. Soc. 95: 7256
72. Fernandez A, Görner H, Kisch H (1985) Chem. Ber. 118: 1936
73. Lahner S, Wakatsuki Y, Kisch H (1987) Chem. Ber. 120: 1011
74. Megehee EG, Johnson CE, Eisenberg R (1989) Inorg. Chem. 28: 2423
75. Linhard M (1944) Z. Elektrochem. 50: 224
76. Linhard M, Weigel M (1951) Z. Anorg. Allg. Chem. 266: 49
77. Schmidtke H-H (1963) Z. Phys. Chem. N.F.: 38: 170
78. Pina F, Ciano M, Mulazzani QG, Venturi M, Balzani V, Moggi L (1984) Sci. Papers I.P.C.R. 78: 166
79. Pina F, Ciano M, Buggi L, Balzani V (1985) Inorg. Chem. 24: 844
80. Pina F, Mulazzani QG, Venturi M, Ciano M, Balzani V (1985) Inorg. Chem. 24: 848
81. Sugimoto H, Hataoka H, Mori M (1982) J. Chem. Soc. Chem. Commun.: 1301
82. Waysbort D, Evenor M, Navon G (1975) Inorg. Chem. 14: 514
83. Sexton DA, Curtis JC, Cohen H, Ford PC (1984) Inorg. Chem. 23: 49
84. Elsbernd H, Beattie JK (1968) Inorg. Chem. 7: 2468
85. Clark SF, Watts RJ, Dubois DL, Connolly JS, Smart JC (1985) Coord. Chem. Rev. 64: 273
86. Orgel LE (1954) Quarterly Rev. 8: 422
87. Harris CM, McKenzie ED (1963) J. Inorg. Nucl. Chem. 25: 171
88. Matsubara T, Efrima S, Metiu HT, Ford PC (1979) J. Chem. Soc. Faraday Trans II, 75: 390
89. Kalisky O, Shirom M (1977) J. Photochem. 7: 215
90. Vogler A, Losse W, Kunkely H (1979) J. Chem. Soc. Chem. Commun.: 187
91. Brand JCD, Snedden W (1957) Trans. Faraday Soc. 53: 894
92. Traverso O, Scandola F (1970) Inorg. Chim. Acta 4: 493
93. Bock CR, Wrighton MS (1977) Inorg. Chem. 16: 1309
94. Fitch IW, Lagowski JJ (1966) J. Organomet. Chem. 5: 480
95. Huttner G, Fischer EO, Fischer RD, Carter OL, McPhail AT, Sim GA (1966) J. Organomet. Chem. 6: 288
96. Huttner G, Fischer EO (1967) J. Organomet. Chem. 8: 299
97. Kobayashi H, Kobayashi M, Kaizu Y (1973) Bull. Chem. Soc. Jpn. 46: 3109
98. Kobayashi H, Kobayashi M, Kaizu Y (1975) Bull. Chem. Soc. Jpn 48: 1222
99. Holland GF, Mannig MC, Ellis DE, Trogler WC (1983) J. Am. Chem. Soc. 105: 2308
100. Lilie J, Shinohara N, Simic MG (1976) J. Am. Chem. Soc. 98: 6516
101. Shinohara N, Lilie J, Simic MG (1977) Inorg. Chem. 16: 2809
102. Haim A, Wilmarth WK (1961) J. Am. Chem. Soc. 83: 509
103. Vogler A, Kunkely H (1975) Ber. Bunsenges. Phys. Chem. 79: 301
104. Vogler A, Osman AH, Kunkely H (1987) Inorg. Chem. 26: 2337
105. Larsson R (1967) Acta Chem. Scand. 21: 257
106. Kane-Maguire NAP, Langford CH (1973) J. Chem. Soc. Chem. Commun.: 351
107. Langford CH, Sasseville RLP (1981) Can. J. Chem. 59: 647
108. Creutz C, Kroger P, Matsubara T, Netzel TL, Sutin N (1979) J. Am. Chem. Soc. 101: 5442
109. Mirbach MF, Mirbach MJ, Wegman RW (1984) Organometallics 3: 900
110. McCleverty JA, Davison A, Wilkinson G (1965) J. Chem. Soc.: 3890
111. Meyer TJ, Caspar JV (1985) Chem. Rev. 85: 187

112. Tyler DR (1988) Progr. Inorg. Chem. 36: 125
113. Doyle MP, Guy JK, Brown KC, Mahapatro SN, VanZyl CM, Pladziewicz JR (1987) J. Am. Chem. Soc. 109: 1536
114. Adamson AW, Sporer A (1958) J. Am. Chem. Soc. 80: 3865
115. Rehorek D, Schmidt D, Hennig H (1980) Z. Chem. 20: 223
116. Hennig H, Walther D, Thomas P (1983) Z. Chem. 23: 446
117. Nakashima M, Kida S (1982) Bull. Chem. Soc. Jpn. 55: 809
118. Matsubara T, Ford PC (1978) Inorg. Chem. 17: 1747
119. Bockman TM, Kochi JK (1989) J. Am. Chem. Soc. 111: 4669

Metal Complexes as Light Absorption and Light Emission Sensitizers

Vincenzo Balzani, Francesco Barigelletti, and Luisa De Cola

Dipartimento di Chimica "G. Ciamician" dell' Università and Istituto FRAE-CNR, Bologna, Italy

Table of Contents

Topics in Current Chemistry, Vol. 158

1 Introduction

A substantial fraction of the chemical literature is currently dealing with reactions involving light. Electron transfer reactions are particularly important for the connection between chemistry and light because they can be driven by light (photoinduced electron transfer reactions) or can produce light (chemiluminescent electron transfer reactions).

Recent developments in chemical research have shown that several transition metal complexes can be used as "mediators", i.e. as *light absorption sensitizers*, **LAS**, and *light emission sensitizers*, **LES**, in electron transfer reactions involving light. Since these processes are extremely important for theoretical reasons as well as for practical applications (e.g. photochemical conversion of solar energy [1–6]), it seems worthwhile to review the progress made in this field in the last ten years.

Particular emphasis will be placed on the identification of the requirements needed for such mediators and on the possibility of tuning the properties of transition metal complexes in order to meet such requirements. Some examples of the use of **LAS** and **LES** will be discussed, including processes of applicative scope (e.g. photochemical conversion of solar energy) and of curious significance (e.g. oscillating chemiluminescence).

2 Electron Transfer Reactions Involving Light

In electron transfer processes light can participate as a reactant, Eq. (1) or it can be generated as a product, Eq. (2). In both

$$A + h\nu \rightarrow {}^*A \quad (1a)$$

$$^*A + B \rightarrow A^+ + B^- \quad (1b)$$

cases the involvement of light takes place via the formation of

$$A^+ + B^- \rightarrow {}^*A + B \quad (2a)$$

$$^*A \rightarrow A + h\nu \quad (2b)$$

electronically excited states, which are usually denoted by an asterisk.

Electron transfer reactions involving thermally equilibrated excited states can be discussed on the basis of the same thermodynamic and kinetic arguments used to discuss electron transfer reactions of ground states species [7]. To a first approximation, the reduction and oxidation potentials of an excited state are given by Eqs. (3) and (4), where $E^0(A/A^-)$ and $E^0(A/A^+)$ are the reduction and

$$E^0({}^*A/A^-) = E^0(A/A^-) + E({}^*A) \quad (3)$$

$$E^0({}^*A/A^+) = E^0(A/A^+) + E({}^*A) \quad (4)$$

oxidation potentials of the ground state molecule and E(*A) is the one-electron potential corresponding to the zero-zero spectroscopic energy of the excited state [8]. From Eqs. (3) and (4) it follows that an excited state is *both* a better oxidant *and* a better reductant compared to the ground state. It should be noted, however, that the participation of excited states as reactants in electron transfer processes may be prevented by their short lifetimes (vide infra).

2.1 Light as a Reactant

When light is used as a reactant, Eq. (1), there are two possible energetic situations schematically represented in Fig. 1 [9]. The scheme of Fig. 1a corresponds to an

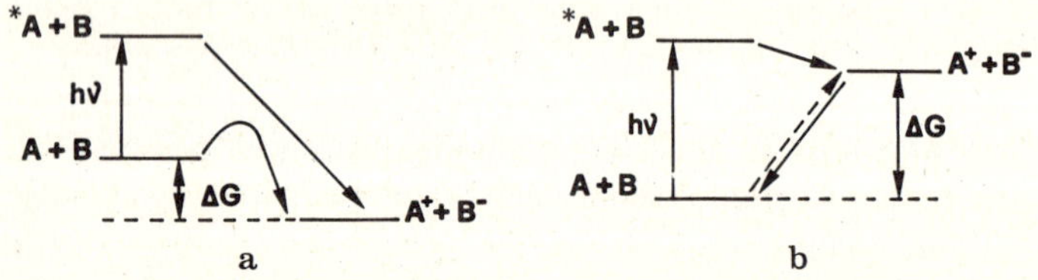

Fig. 1. Schematic representation of the two possible energetic situations for electron transfer reactions involving an excited state reactant (see text)

exergonic dark reaction which is slow for kinetic reasons (high activation energy). Upon light excitation, the reductant A is transformed in the much more powerful reductant *A (vide supra), so that the reaction between *A and B is much more exergonic than the reaction between A and B. Since the activation energy generally decreases with increasing exergonicity (at least for not too strongly exergonic processes [10]), the reaction involving the excited state will be much faster than that involving the ground state. In a system of this kind, light plays the role of a catalyst because it is used to overcome a kinetic barrier. The scheme shown in Fig. 1b corresponds to the case of a dark reaction that cannot take place because of thermodynamic reasons. Light excitation again causes the formation of *A which is a reductant much stronger than A, so that the excited state reaction is thermodynamically allowed. Light can thus drive A + B to $A^+ + B^-$ *via* *A + B, and in such a process, a fraction of the light energy is converted into chemical energy of the products. The converted energy is then released when A^+ and B^- undergo the back electron transfer reaction leading to A + B.

It should be pointed out that usually the number of electrons transferred in these systems is much smaller that the number of absorbed photons for two reasons: (i) only a fraction of the absorbed photons leads to the potentially reactive excited state, and (ii) only a fraction of the molecules that reach the potentially reactive excited state can react because of problems related to the excited state lifetime (τ). For a bimolecular excited state process, Eq. (5), the competition between excited state decay and excited state reaction is regulated

$$\begin{array}{l} {}^*A + B \xrightarrow{k_q} A^+ + B^- \\ \downarrow {\scriptstyle 1/\tau} \\ A \end{array} \tag{5}$$

by the values of $1/\tau$ and $k_q[B]$. For an intramolecular electron transfer process, Eq. (6), the term involved in the competition are $1/\tau$ and k_i.

$$\begin{array}{l} {}^*A{-}B \xrightarrow{k_i} A^+{-}B^- \\ \downarrow {\scriptstyle 1/\tau} \\ A{-}B \end{array} \tag{6}$$

2.2 Light as a Product

The energetic situation of electron transfer processes generating light, Eq. (2), can be schematically represented [9] as in Fig. 2 which differs from the scheme shown in Fig. 1b because the energy content of $A^+ + B^-$ is higher than that of $^*A + B$. In such a case the reaction from $A + B$ to $A^+ + B^-$ can be driven neither thermally nor photochemically. When the oxidant A^+ and the reductant B^- can be prepared in some other ways (e.g. electrochemically) and are mixed together, their electron transfer reaction can lead either to $A + B$, with complete dissipation of the excess free energy into heat, or to $^*A + B$, with dissipation of a smaller amount of free energy. In the latter case, radiative deactivation of $^*A$ leads to generation of light. Under these conditions a fraction of the chemical energy available to the reactants is converted into light energy, the reverse of what occurs in the case illustrated by Fig. 1b. The processes schematized by Figs. 1 and 2 can really be viewed as a chemical equilibrium, Eq. (7) which is displaced from left to right or from right to left depending on whether the energetic situation

$$A + B + h\nu \rightleftarrows {}^*A + B \rightleftarrows A^+ + B^- \tag{7}$$

of the system is depicted by Fig. 1 or 2. When $^*A + B$ and $A^+ + B^-$ are almost isoergonic as in the case of A = $Cr(4{,}7\text{-}Me_2\text{-}phen)_3^{3+}$ and B = $Ru(bpy)_3^{2+}$, both processes can be studied for the same system [11].

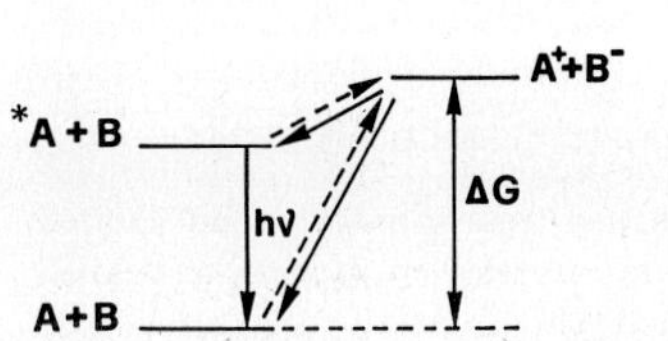

Fig. 2. Schematic representation of the energetic situation for an electron transfer chemiluminescent reaction

Formation of an electronically excited state is, of course, a necessary but not sufficient condition for light generation. Several excited states, in fact, are not luminescent under the experimental conditions used for electron transfer reactions. More generally, the number of generated photons is much lower than the number of transferred electrons for two reasons: (i) the electron transfer process can follow reaction paths that do not lead to the generation of the luminescent excited state; (ii) the luminescence quantum yield of the excited state product is lower than unity.

2.3 Fundamental Concepts on Excited States

Since electron tranfer reactions involving light (as a reactant or as a product) take place *via* formation of electronically excited states, we need to recall some fundamental concepts on the excited state behavior.

Figure 3 shows a schematic Jablonski diagram for a typical molecule. Light absorption by the ground state molecule, A, leads mostly to the spin-allowed excited states $A(a_1)$, $A(a_2)$, $A(a_3)$, etc. The spin-forbidden excited states, $A(f_1)$, $A(f_2)$, $A(f_3)$, etc., can be populated by intersystem crossing (isc) from the corresponding spin allowed states. Deactivation of upper lying excited states to the lowest one of the same multiplicity is generally very fast (picosecond time scale), so that only the lowest excited state of each multiplicity, $A(a_1)$ and $A(f_1)$, can be expected to play the role of excited state reactant in electron transfer processes. The excited state lifetimes of $A(a_1)$ and $A(f_1)$ are given by Eqs. (8) and (9), where k_f, k_{ic}, k_{isc}, k_p, and k'_{isc} are the rate constants for

$$\tau_{A(a_1)} = 1/(k_f + k_{ic} + k_{isc}) \tag{8}$$

$$\tau_{A(f_1)} = 1/(k_p + k'_{isc}) \tag{9}$$

fluorescence, internal conversion, $A(a_1) \rightsquigarrow A(f_1)$ isc, phosphorescence, and $A(f_1) \rightsquigarrow A$ isc, respectively. For typical organic molecules, $\tau_{A(a_1)}$ is in the nano-

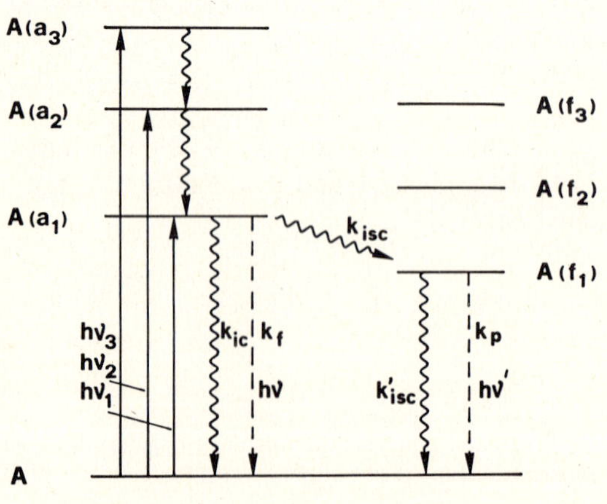

Fig. 3. Jablonski diagram for a typical molecule A; *a* and *f* indicate spin-allowed and spin-forbidden excited states respectively (for more details, see text)

second time scale. In such a case a bimolecular excited state process, Eq. (5), can compete with the excited state decay if $k_q[B] \simeq 10^9\ s^{-1}$. Since the upper limiting value of k_q is the diffusion constant k_d, which is about $10^{10}\ M^{-1}\ s^{-1}$ for the most common solvents at room temperature, substantial concentrations of the reaction partner B are required for Eq. (5) to occur. By contrast, the lowest spin-forbidden excited state $A(f_1)$ of organic molecules is very long-lived (second to millisecond time scale) and has ample opportunity to participate in electron transfer processes. In practice, however, such an excited state can hardly be kept under control because it can undergo self-quenching processes and quenching by impurities [12, 13]. For transition metal complexes the situation is quite different because of the presence of the (heavy) metal atom causes a relatively strong spin-orbit coupling [14]. As a consequence, the formally spin-forbidden processes $A(a_1) \rightsquigarrow A(f_1)$ and $A(f_1) \rightsquigarrow A$ become much faster. This brings $\tau_{A(a_1)}$ in the subnanosecond time range, thereby preventing $A(a_1)$ to participate in bimolecular electron transfer processes, and increases the efficiency of the $A(a_1) \rightsquigarrow A(f_1)$ intersystem crossing (often, up to unity). The excited state lifetime of $A(f_1)$ also becomes shorter, but in several cases it remains in the 10^{-5} to 10^{-8} s time scale, which is sufficiently short to prevent uncontrolled quenching processes and sufficiently long to allow the involvement of $A(f_1)$ in electron transfer processes [8].

In the case of organic molecules, only the lowest spin-allowed excited state $A(a_1)$ is usually luminescent, but direct formation of such a high energy excited state in a chemical reaction requires a strongly exergonic process; when $A(f_1)$ is formed, luminescence is not observed unless subsequent triplet-triplet annihilation processes lead back to $A(a_1)$ [15]. For coordination compounds, the luminescent level is the lowest spin forbidden excited state, $A(f_1)$, which usually lies at relatively low energy, so that even moderately exergonic reactions can lead to light emission [16].

2.4 Electron Transfer Kinetics

The theoretical treatment of the kinetics of electron transfer reactions (both as bimolecular and unimolecular processes) has been one of the major topics in physical chemistry for the last twenty years and is reviewed in a number of excellent reviews [10, 17–21]. In this section we will simply recall, within the framework of a classical nonadiabatic approach [10, 19], the various factors which are expected to affect the rate of unimolecular electron transfer processes. More details and a deeper discussion are presented in other chapters of this series.

The unimolecular rate constant of an electron transfer reaction, Eq. (10), is given by Eq. (11). The quantities involved in this equation can be discussed by

$$A \ldots B \rightarrow A^- \ldots B^+ \tag{10}$$

$$k_e = \nu_n \kappa \exp(-\Delta G^{+}/RT) \tag{11}$$

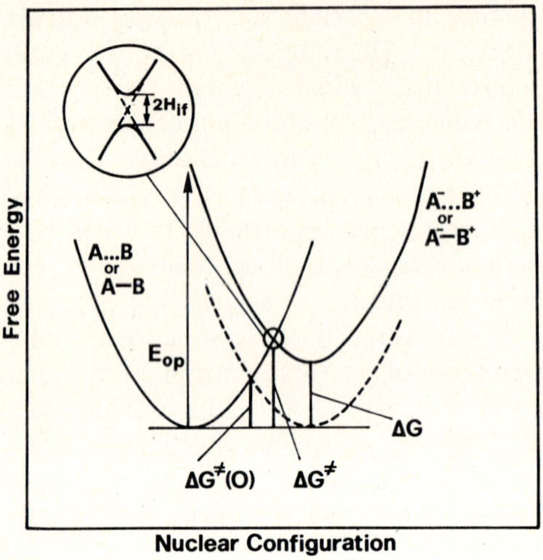

Fig. 4. Schematic representation of thermal and photoinduced electron transfer processes. (For details, see text)

referring to Fig. 4, which depicts the electron transfer process in terms of free energy surfaces for reactants and products.

As far as the exponential term of Eq. (11) is concerned, the free energy of activation $\Delta G^{\neq}$ is related to the energy required to reach the distorted nuclear configuration (crossing point) at which the Franck-Condon principle is satisfied for the electron transfer process. This energy is classically related, by means of a free energy relationship, Eq. (12), to an "intrinsic barrier" parameter, $\Delta G^{\neq}(0)$,

$$\Delta G^{\neq} = \Delta G^{\neq}(0)\,[1 + \Delta G/4\,\Delta G^{\neq}(0)]^2 \tag{12}$$

which represents the degree of distortion (horizontal displacement in Fig. 4) between the reactant and product curves, and the actual free energy change ΔG (vertical displacement in Fig. 4) of the electron transfer process. The intrinsic barrier $\Delta G^{\neq}(0)$ can be viewed as the sum of distinct contributions [10] from solvent reorganization, $\Delta G^{\neq}(0)_{out}$, and reorganization in the inner coordination sphere, $\Delta G^{\neq}(0)_{in}$:

$$\Delta G^{\neq}(0) = \Delta G^{\neq}(0)_{out} + \Delta G^{\neq}(0)_{in}\,. \tag{13}$$

Appropriate equations are available to evaluate the outer and inner contributions in terms of molecular parameters and solvent dielectric properties [10, 19]. For a given degree of distortion, the classical model predicts that $\Delta G^{\neq}$ initially decreases with increasing exergonicity, goes to zero for $\Delta G = -4\,\Delta G^{\neq}(0)$, and then increases again for larger exergonicities. This last region is called the "Marcus inverted" region of electron transfer reactions [17]. According to more rigorous, but less practical, quantum mechanical models [22–24], an inverted behavior (rate decreasing with increasing driving force) is still predicted to occur, although with some quantitative differences with respect to the classical model (linear instead of quadratic dependence, usually referred to as "energy gap law" [25]). Because of problems related to diffusion control and other reasons which are not yet completely clear, the inverted behavior often escaped experimental detection in bimolecular electron transfer reactions [26, 27], leading to the use of empirical asymptotic free energy relationships [28, 29]. On the other hand, electron transfer processes in rigid matrixes [30], geminate radical ion pairs [31, 32], or covalently bound supramolecular systems [33–35], offer clear examples of the inverted behavior. The energy gap law is also obeyed for a number of electronically excited states whose decay can be viewed as an intramolecular electron transfer [36, 37].

The v_n term in the pre-exponential part of the rate constant, Eq. (11), is an effective nuclear frequency for motion on the parabolic energy surfaces of Fig. 4 and can be expressed as a function of both the outer solvent reorganizational frequency, v_{out}, and inner vibrational frequencies of the reactant and product species, v_{in} [10, 26]:

$$v_n^2 = \frac{v_{out}^2 \Delta G^{\ddagger}(0)_{out} + v_{in}^2 \Delta G^{\ddagger}(0)_{in}}{\Delta G^{\ddagger}(0)_{out} + \Delta G^{\ddagger}(0)_{in}} . \tag{14}$$

When $\Delta G^{\ddagger}(0)_{out} \gg \Delta G^{\ddagger}(0)_{in}$, v_n is dominated by the solvent reorganizational frequency. Examples of electron transfer reactions that are controlled by the dielectric relaxation time of the solvent have been reported (see, for example, [38, 39]).

The κ term in the pre-exponential part of the rate constant, Eq. (11), is the transmission coefficient, related to the probability of conversion from reactant to product in the avoided crossing region (Fig. 4), which can be expressed as a function of the electronic coupling matrix element H_{if} between the initial and final states of the systems:

$$\kappa = \frac{2[1 - \exp(-v_{el}/2v_n)]}{2 - \exp(-v_{el}/2v_n)} , \tag{15}$$

$$v_{el} = 2H_{if}^2/h[\pi^3/4\, \Delta G^{\ddagger}(0)RT]^{1/2} . \tag{16}$$

The electronic interaction H_{if} depends essentially on the overlap between the donor and acceptor orbitals [40, 41]. In bimolecular electron transfer processes, the geometric configuration of the encounter is not fixed, and a number of different configurations are explored within the lifetime of the encounter complex. This may lead to some kind of optimization of the transmission coefficient, which is actually considered to be unitary for many bimolecular electron transfer reactions (adiabatic behavior). In intramolecular processes, such as those involving ion-pairs or supermolecules, the geometry is fixed or at least restricted and the actual magnitude of H_{if} can play a more important role. It should be noted that according to Eqs. (15) and (16) $\kappa \simeq v_{el}/v_n$ for small H_{if} and $\kappa \simeq 1$ when H_{if} is large. Thus, the preexponential factor is proportional to $(H_{if})^2$ for small H_{if} (non-adiabatic regime) and saturates to v_n for large H_{if} (adiabatic regime).

3 Mediators of Electron Transfer Processes

3.1 Thermal Reactions: Relays

It is well known that very slow thermal electron transfer reactions (including electrodic processes) can become faster in the presence of suitable electron carriers, usually called relays (R) (Fig. 5). A simple example is that concerning the reduction of $Co(NH_3)_6^{3+}$ by Eu_{aq}^{2+} ions, which can be accelerated by pyridine derivatives [42]. The relay takes place in the process but it is not consumed. Of course, a relay cannot have any effect on thermodynamically forbidden reactions.

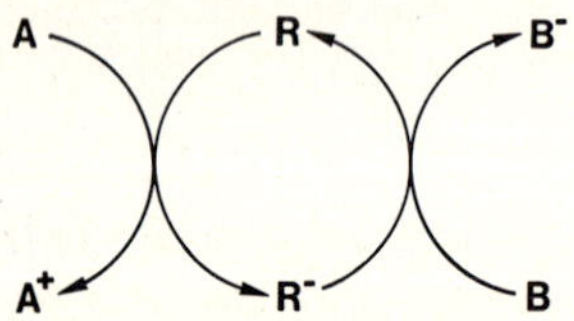

$$A + B \xrightarrow{R} A^+ + B^-$$

Fig. 5. Thermal electron transfer reaction mediated by a relay R

3.2 Photoinduced Reactions: Light Absorption Sensitizers (LAS)

As we have seen in Sect. 2.1, light energy can be used to induce thermodynamically allowed (Fig. 1a) or forbidden (Fig. 1b) electron transfer reactions. In principle, any electron transfer process can be photoinduced, but necessary conditions are (i) absorption of light with formation of an electronically excited state which exhibits (ii) a sufficient energy content to make the excited state reaction thermodynamically allowed and (iii) a lifetime long enough to allow the reaction to compete with excited state decay.

Several electron transfer processes do not satisfy such conditions. Consider, for example, the reduction of methylviologen by triphenylamine, Eq. (17) [43]. Such a reaction cannot occur in the dark because it is endergonic by 1.45 eV.

$$MV^{2+} + NPh_3 \rightarrow MV^+ + NPh_3^+ \quad (17)$$

In principle, it could be driven by visible light, Eq. (18), since visible photons have an energy content ranging from 1.56 eV (795 nm) to 3.10 eV (400 nm).

$$MV^{2+} + NPh_3 + \text{light (visible)} \rightarrow MV^+ + NPh_3^+ \quad (18)$$

However, neither MV^{2+} nor NPh_3 absorb in the visible, so that Eq. (18) does not occur. This is also the case for the very important water splitting reaction (Sect. 6.1) which, in principle, could be driven by visible light but, in practice, cannot occur directly because water is transparent in the visible region.

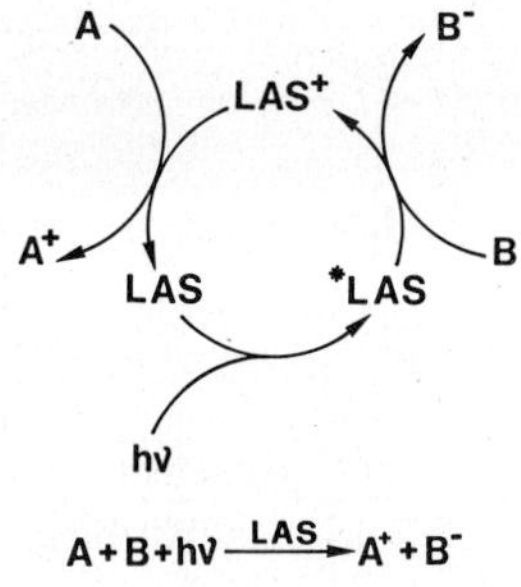

Fig. 6. Light driven electron transfer process *via* the intermediacy of a light absorption sensitizer (LAS)

Such "potential" photochemical reactions can be mediated (or sensitized) by species which exhibit specific spectroscopic, redox, and excited state properties. For example, the photoreduction of methylviologen by triphenylamine [43] takes place in the presence of $Ru(bpy)_3^{2+}$, Eq. (19).

$$MV^{2+} + NPh_3 + \text{light(visible)} \xrightarrow{Ru(bpy)_3^{2+}} MV^{+} + NPh_3^{+} \tag{19}$$

The task of the mediator is to absorb light and to convert light energy into chemical energy available to the electron transfer process, without being consumed. Such mediators are usually called photosensitizers. We prefer to call them *Light Absorption Sensitizers* **(LAS)** [9] to underline their role and to use a parallel expression in the case of the mediators of chemiluminescent electron transfer processes (vide infra).

Figure 6 shows schematically the role played by a **LAS** in a photoinduced electron transfer process: (i) it must absorb light so as to give an excited state; (ii) this excited state must be able to reduce (or oxidize) one of the reactants; (iii) the oxidized (or reduced) form of the **LAS** must be able to complete the redox process and to regenerate the ground state **LAS.** It must be pointed out that a **LAS** always converts light energy into free chemical energy and that such converted energy can be either completely dissipated during the overall process (when the dark reaction is exergonic), or partially found as free energy of the products (for endergonic dark reactions).

Usually, the number of electrons transferred via a **LAS** is much smaller than the number of absorbed photons, for the same reasons previously discussed in the case of direct photoinduced processes.

3.3 Chemiluminescent Reactions: Light Emission Sensitizers (LES)

As we have seen in Sect. 2.2, strongly exergonic electron transfer reactions can give rise to light emission (Fig. 2). Necessary conditions are (i) high exergonicity (ii) to form an electronically excited state (iii) which must be luminescent.

Several electron transfer reactions are sufficiently exergonic to produce light, but do not satisfy conditions (ii) or (iii). For example, the reaction between lead dioxide and oxalate, Eq. (20), is sufficiently exergonic (~2 eV) to generate

$$1/2\ PbO_2 + 1/2\ C_2O_4^{2-} + 2H^+ \rightarrow 1/2\ Pb^{2+} + CO_2 + H_2O \qquad (20)$$

visible light, but it only produces heat because electronic excited states cannot be obtained.

Such "potential" chemiluminescent reactions can be "mediated" (or sensitized), as it happens for the "potential" photochemical reactions, by species exhibiting specific spectroscopic, redox, and excited state properties. For example, visible light is emitted when the reaction between lead dioxide and oxalate is performed in the presence of $Ru(bpy)_3^{2+}$ [44]:

$$1/2\ PbO_2 + 1/2\ C_2O_4^{2-} + 2\ H^+ \xrightarrow{Ru(bpy)_3^{2+}}$$
$$1/2\ Pb^{2+} + CO_2 + H_2O + \text{light} \qquad (21)$$

In these processes the task of the mediator is to make use of the free energy of the reaction to produce light, without being consumed. In other words, the mediator plays the role of a sensitizer for light generation, and can thus be called *Light Emission Sensitizer* (**LES**). Figure 7 shows schematically the role played by a **LES** in a mediated chemiluminescent process: (i) it must be oxidized (or reduced) by one of the reactants; (ii) its oxidized (or reduced) form must then oxidize (or reduce) the other reactant to yield an excited state, ***LES**; (iii) such an excited state must undergo radiative deactivation (luminescence), regenerating the ground state **LES.**

In most cases the photons generated by a **LES** are only a small fraction of the electrons transferred in the overall electron transfer reaction. This is due not only to the lower than unity luminescence quantum yield of ***LES**, but also to the fact that **LES** mediates only a fraction of the overall electron transfer events. It should also be noticed that, as it may happen for direct chemiluminescent process, in a complex electron transfer process of overall low exergonicity some light emission can still occur because a side reaction or a minor reaction path may be sufficiently exergonic to produce the luminescent excited state of **LES.**

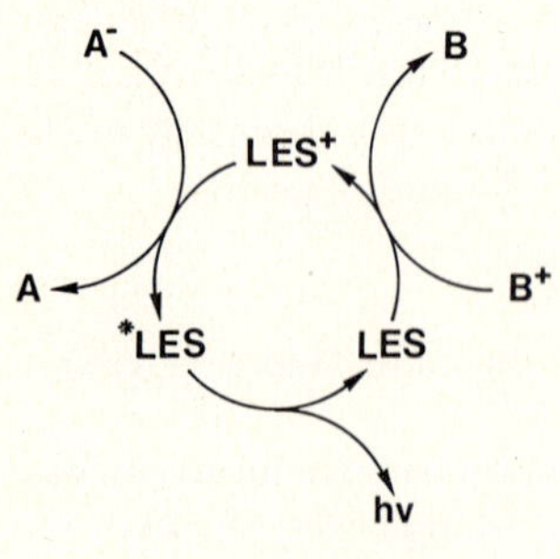

$$A^- + B^+ \xrightarrow{LES} A + B + h\nu$$

Fig. 7. Light generating electron transfer process *via* the intermediacy of a light emission sensitizer (LES)

3.4 Requirements Needed for LAS and LES

A **LAS** is a chemical species that must be able to use light energy to induce an electron transfer reaction between two substrate species (Fig. 6). To play this role, a **LAS** must be involved in excitation and redox processes without being consumed. The requirements needed for an ideal **LAS** can be listed making reference to Fig. 8a:

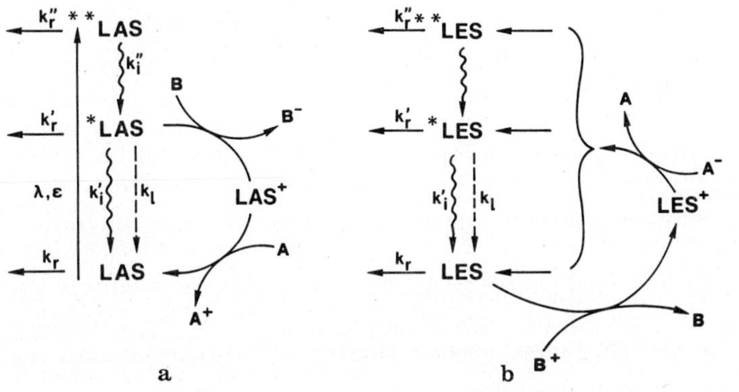

Fig. 8a, b. Schematic representation of the processes taking place in, and the properties needed for, LAS (**a**), and LES (**b**). Subscripts *r*, *i*, and *l* indicate chemical reactions, radiationless decay, and luminescence, respectively

(i) stability in the ground state, in the excited states, and in the oxidized (or reduced) form;
(ii) absorption of light in a suitable spectral region;
(iii) high efficiency of population of the excited state which is responsible for electron transfer;
(iv) suitable ground and excited state redox potentials;
(v) reasonably long excited state lifetime;
(vi) high efficiency of the redox processes with the substrates.

A **LES** is a chemical species that must be able to use the chemical energy of a redox reaction to be promoted to a luminescent excited state (Fig. 7). A **LES** must thus be involved in redox and luminescence processes without being consumed. The requirements needed for an ideal **LES** can be listed making reference to Fig. 8b:

(i) stability in the ground state, in the excited state(s), and in the oxidized (or reduced) form;
(ii) suitable redox potentials in the ground and excited states;
(iii) high efficiency of the redox process which leads to the luminescent excited state;
(iv) suitable excited state energy;
(v) high quantum yield of luminescence.

From the above lists it is clear that the stability and redox requirements are the same for **LAS** and **LES**. From a spectroscopic point of view, the requirements are somewhat different (e.g. a **LAS** does not need to exhibit luminescence and a **LES** does not need to have intense absorption bands), but slow radiationless deactivation of the active excited state is a fundamental requirement in both cases because a **LAS** must have a long excited state lifetime and a **LES** must have a high luminiscence efficiency. Therefore it is not surprising that a compound which plays the role of **LAS** can often play the role of **LES**, and vice versa.

3.5 Why Metal Complexes?

It is extremely difficult to find molecules that satisfy the requirements needed for **LAS** and **LES**. Metal complexes containing aromatic ligands are good candidates for the following reasons, some of which have already been mentioned in Sect. 2.3:

a) They exhibit metal-to-ligand charge transfer (MLCT) bands at relatively low energies (near u.v. and visible spectral regions); this is a fundamental property for a **LAS**. Of course, neither simple (aquated) metal ions nor organic molecules can show such bands.
b) Because of the presence of the (heavy) metal atom, metal complexes exhibit noticeable spin-orbit coupling so that the formally spin-forbidden $A(a_1) \rightsquigarrow A(f_1)$ process of Fig. 3 (****LAS** ⇝ ***LAS** in Fig. 8a) becomes very fast and the efficiency of population of the lowest excited state is often unity. This is a quite important property for a **LAS**, usually not satisfied by organic molecules.
c) Spin-orbit coupling increases the decay rate of the lowest excited state to the ground state. This often brings the lifetime of the lowest excited state in the microsecond time region, which is a very convenient one for a **LAS**. For simple metal ions the excited state lifetime is usually extremely short (except for intraconfigurational excited states). For organic molecules the lifetime of the lowest spin-allowed excited state is generally very short, and that of the lowest spin-forbidden excited state is often so long as to allow the occurrence of self-quenching or quenching by impurities.
d) Because of the presence of the metal ion and ligands, metal complexes possess distinct redox centers and can more easily exhibit the redox properties needed for **LAS** or **LES**.
e) The lowest excited state is often luminescent in fluid solution at room temperature. This property is essential for **LES** and quite useful for **LAS**. In the field of organic molecules, fluid-solution room-temperature luminescence is usually exhibited by the lowest spin-allowed excited state which is very short lived. In the case of simple metal ions, luminescence in fluid solution at room temperature is only observed when the lowest excited state is intraconfigurational.
f) Most of the ground and excited state properties may be tuned by a judicious choice of the metal and/or ligands [14, 45].

4 Metal Complexes

In the discussion that follows we will examine the properties of the "species" involved when a complex plays the role of **LAS** or **LES** (Fig. 8): the ground state molecule, the one-electron oxidized and/or reduced forms, and the lowest excited states.

4.1 Localized Molecular Orbital Approximation

Coordination compounds are made of metal ions and ligands which can also exist separately from each other. For this reason, as well as for the sake of convenience, the kinetic, spectroscopic, and redox properties of coordination compounds are usually discussed with the assumption that the ground state, the excited states, and the redox species can be described in a sufficiently approximate way by localized molecular orbital configurations. With such an assumption, the various electronic transitions are classified as metal centered, MC, ligand centered, LC, and charge transfer, CT, (either metal-to-ligand, MLCT, or ligand-to-metal, LMCT) [46, 47], and the oxidation and reduction processes are classified as metal- or ligand-centered [48–50]. This simplified picture, of course, is no longer applicable if there is a large degree of covalency in the metal-ligand bonds and if the excited configurations of different orbital nature are sufficiently close in energy to be intermixed.

4.2 Ground State

Most coordination compounds are labile in the ground state, i.e. they undergo ligand substitution reactions, especially in coordinating solvents like water. This is a severe drawback because the spectroscopic and redox properties of a complex are critically related to the composition and structure of the first coordination sphere. Once coordinated to the metal ion, the solvent molecules favor the radiationless deactivation of the excited states to the ground state *via* state coupling by high energy phonons. Furthermore, the weak ligand field of the solvent molecules results in low-lying, displaced excited states [46, 47] which are strongly coupled to the ground state [25, 51]. For this reasons, complexes containing solvent molecules in their first coordination sphere are useless as **LAS** or **LES**. Among classical coordination compounds, only the d^3 and d^6 octahedral complexes and the d^8 square planar complexes are usually inert in the ground state [52]. The complexes of lanthanide ions are very labile and in most cases they give rise to equilibria involving several species that contain a variable number of solvent molecules in the first coordination sphere [53].

A way to remedy the drawback of chemical lability is to enclose the metal ion into a suitable cage ligand. Intermediate steps along this direction are constituted by the use of chelating or macrocyclic ligands (Fig. 9). As we will see later, such an encapsulation strategy can also be employed to protect a complex towards photodissociation and to avoid disruption of the molecular structure

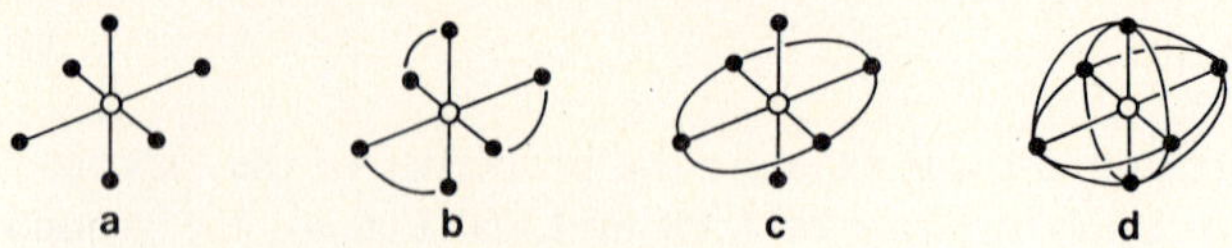

Fig. 9a–d. Schematic representation of complexes containing (a) monodentate, (b) chelate, (c) macrocyclic, and (d) cage ligands

when the oxidation state of the metal is modified by a redox process. The best example of "forced" ground state inertness by encapsulation of the metal is that concerning the lanthanide ions. Contrary to what happens for the related complex of monodentate ligands, the lanthanide cryptates of the 2.2.1 [54] and bpy.bpy.bpy [55] ligands (Fig. 10) have well defined coordination spheres and

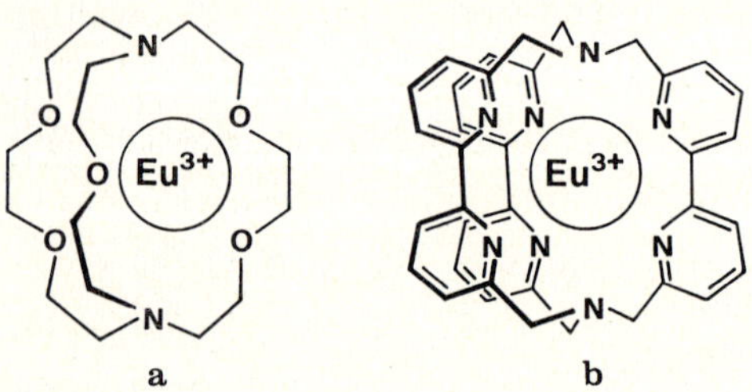

Fig. 10a, b. Structural formulae of the 2.2.1. (a), and bpy.bpy.bpy. (b) cryptates of europium ions

are kinetically inert. The advantages offered by these cage-type complexes compared to the lanthanide aquo ions, as far as the luminescent properties are concerned, are discussed in detail elsewhere [56]. As to their possible use as **LAS** and **LES**, it should be noted that among the lanthanide ions only Eu^{3+} undergoes a redox process in an useful potential range.

Cage-type ligands, however, are difficult to prepare and even more difficult is the encapsulation of the desired metal ion into such cages. For this reason, only a few examples of cage type complexes of photochemical and photophysical interest are known [56–58] and investigations concerning the search of **LAS** and **LES** are usually restricted to the "naturally" inert d^3 octahedral Cr(III) complexes, d^6 octahedral Fe(II), Co(III), Ru(II), Rh(III), Re(I), Os(II), and Ir(III) complexes, and d^8 square planar Pt(II) complexes.

4.3 Oxidized and Reduced Species

One-electron metal-centered oxidation of Cr(III) complexes can hardly be observed. One-electron metal-centered oxidation of d^6 octahedral complexes is not expected to cause ligand dissociation because the outgoing electron is taken from an essentially nonbonding π metal orbital and d^5 octahedral complexes are known [52]. One-electron metal-centered oxidation of d^8 square planar complexes is

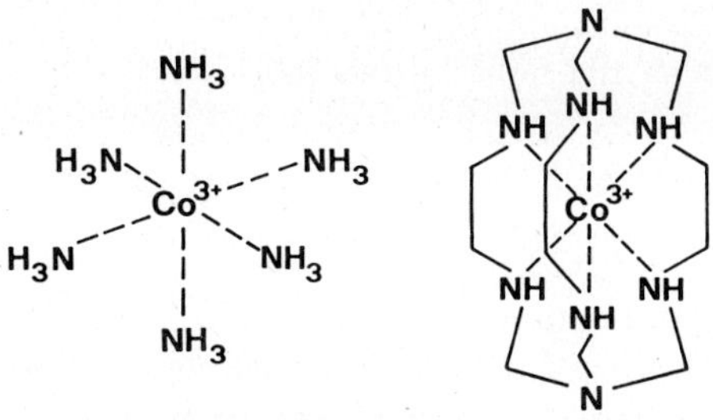

a b

Fig. 11a, b. Molecular structure of hexamine (**a**) and sepulchrate (**b**) Co(III) complexes

usually followed by chemical reaction since d^7 complexes are not stable. One-electron metal-centered reduction of d^3, d^6, and d^8 complexes is expected to give ligand dissociation since the incoming electron goes in a σ^* M-L antibonding orbital (possible exceptions are very strong field d^3 complexes). The instability of the redox species is, of course, a severe drawback when a high turnover number is required for a **LAS** or **LES** system. Ligand dissociation on oxidation and/or reduction can be prevented by using cage-type ligands. This strategy is best illustrated by the comparison between $Co(NH_3)_6^{3+}$ and its cage analogue, $Co(sep)^{3+}$ (Fig. 11). For the hexamine complex, pulse radiolysis experiments [59] show that one-electron reduction, Eq. (22), is followed by the consecutive release

$$Co^{III}(NH_3)_6^{3+} \xrightarrow{e^-_{aq}} Co^{II}(NH_3)_6^{2+} \tag{22}$$

$$Co^{II}(NH_3)_6^{2+} \xrightarrow[k>1\times10^3 s^{-1}]{H_3O^+} Co^{2+}_{aq} + 6\,NH_4^+ \tag{23}$$

of all the ammonia ligands in the millisecond time scale, Eq. (23), whereas one-electron reduction of $Co(sep)^{3+}$, Eq. (24), leads to a fairly inert complex, Eq. (25) [60].

$$Co^{III}(sep)^{3+} \xrightarrow{e^-_{aq}} Co^{II}(sep)^{2+} \tag{24}$$

$$Co^{II}(sep)^{2+} \xrightarrow[k<10^{-6} s^{-1}]{H_3O^+} Co^{2+} + \text{products} \tag{25}$$

When the redox process concerns a ligand, the resulting complex may or may not be stable depending on the particular nature of the ligand. A quite common and interesting case is that of complexes containing aromatic-type ligands like 2,2′-bipyridine. Such complexes undergo several successive reversible reduction processes which, to a first approximation, can be considered as localized in a single ligand [48, 49]. For example, $Ru(bpy)_3^{2+}$ in DMF solution undergoes six successive reduction steps: the first three steps correspond to successive first one-electron reduction of the three bpy ligands, and the following three steps to successive second one-electron reduction of the same ligands [61].

4.4 Excited States

As mentioned in Sect. 4.1, in simple coordination compounds (which may be exemplified by M-L) one can distinguish the following types of excited states on the basis of a localized MO approach: *metal-centered* (MC), *ligand-centered* (LC), *ligand-to-metal charge transfer* (LMCT), and *metal-to-ligand charge transfer* (MLCT).

As it was emphasized above, for transition metal complexes only the lowest excited state (or the lowest manifold of thermally equilibrated excited states [14]) has a chance of living long enough to participate in redox processes and/or to luminesce. It is, therefore, of paramount importance to know which factors determine the energy of the various types of excited states, and to understand how such factors can be manipulated so as to design complexes having long lived and/or luminescent excited states.

The MC excited states may be intra- or interconfigurational. The first case applies to the lowest MC excited state of octahedral d^3 metal complexes (e.g. Cr(III) complexes). Intraconfigurational excited states have nuclear coordinates almost identical to those of the ground state [46, 47]. Since this disfavors radiationless decay [51], intraconfigurational excited states can be long-lived and can exhibit luminescence. Luminescence bands arising from intraconfigurational MC excited states are usually structured and exhibit a small Stokes shift from the corresponding absorption bands. The d^6 octahedral and d^8 square planar complexes can only have interconfigurational MC excited states [47]. Such excited states are obtained by promoting an electron from a π non-bonding orbital to a σ^* antibonding orbital. Therefore, they exhibit different nuclear coordinates compared to the ground state, a situation which favors radiationless decay, reduces the excited state lifetime, and usually prevents luminescence to occur at room temperature [51]. When observed, emission from interconfigurational excited states gives rise to gaussian-shaped bands considerably red shifted from the lowest energy absorption band and devoided of vibrational structure.

For most coordination compounds the LC excited states lie at high energy and therefore they cannot play any direct role in determining **LAS** or **LES** properties. For d^6 Rh(III) and Ir(III) complexes of polypyridine or other aromatic ligands, however, the lowest excited state is often a ^{3}LC level [14, 62]. In such cases, a structured luminescence band can be observed especially at low temperature, quite similar (except for a slight red shift) to that of the free ligand. The excited state lifetime, however, is substantially (e.g. 10^3–10^4 times) shorter than that of a pure ^{3}LC emission, because the presence of the metal ion induces spin-orbit coupling which speeds up the radiative and radiationless deactivations to the ground state.

LMCT excited states may lie at low energy only with very reducing ligands and/or very oxidizing metal ions [46, 47]. This, of course, is not generally the case for d^8 and d^6 metal complexes because of the strong electron density on the metal. Anyway, a d^8 or d^6 metal complex having a LMCT level as the lowest excited state would not exhibit **LAS** or **LES** properties. Such an excited state, in fact, would be strongly distorted compared with the ground state geometry (because of the presence of an electron in a σ^* antibonding orbital), and therefore it would be very short-lived and non-luminescent. For d^3 octahedral complexes the promoted electron could occupy, in principle, a π non-bonding metal orbital. As mentioned above, however, for Cr(III) complexes the lowest excited state is the ^2E_g intraconfigurational MC excited state in all cases.

MLCT levels lie at low energy when the metal can be easily oxidized and the ligand(s) can be easily reduced [46, 47]. The former condition applies to several metal complexes, and particularly to d^6 octahedral complexes of metals in low oxidation states. The latter condition is usually related to the presence of aromatic ligands which possess low-energy empty π^* orbitals. In these cases, the MLCT transition promotes an electron from a practically non-bonding π metal orbital to a delocalized π^* ligand orbital. Such a change in the electronic configuration does not substantially affect the bond lengths so that the MLCT excited states are usually only slightly distorted compared with the ground state geometry. We have already emphasized that such a situation disfavors radiationless decay and allows luminescence to occur. The lowest MLCT excited state has formally a different spin quantum number compared with the ground state, so that deactivation to the ground state is formally spin-forbidden. The direct involvement of the (heavy) metal ion, however, induces a strong spin-orbit coupling. Therefore, the radiative lifetime of ^{3}MLCT excited states (μs time scale) is usually shorter than the radiative lifetime of ^{3}LC excited states (ms time scale).

From the above observations, it follows that the orbital nature of the lowest excited state can be revealed by examination of the luminescence behavior [14, 45, 63, 64] and that, in a series of complexes of the same metal ion, one can determine the orbital nature of the lowest excited state by a suitable choice of the ligands (Sect. 5.1).

5 Ruthenium(II) Polypyridine Complexes

$Ru(bpy)_3^{2+}$ has certainly been one of the molecules most extensively studied and most widely used in research laboratories during the last ten years. A unique combination of chemical stability, redox properties, excited state reactivity, luminescence emission, and excited state lifetime has attracted the attention of many research workers first on this molecule, and then on some hundreds of its derivatives [45, 65–68]. The great interest generated by the study of this class of complexes has stimulated the growth of several branches of chemistry such as photochemistry, photophysics, photocatalysis, electrochemistry, photoelectrochemistry, chemi- and electrochemi-luminescence, and electron and energy transfer. In particular, Ru(II)-polypyridine complexes exhibit excellent properties to play the role of **LAS** and **LES** (Fig. 12).

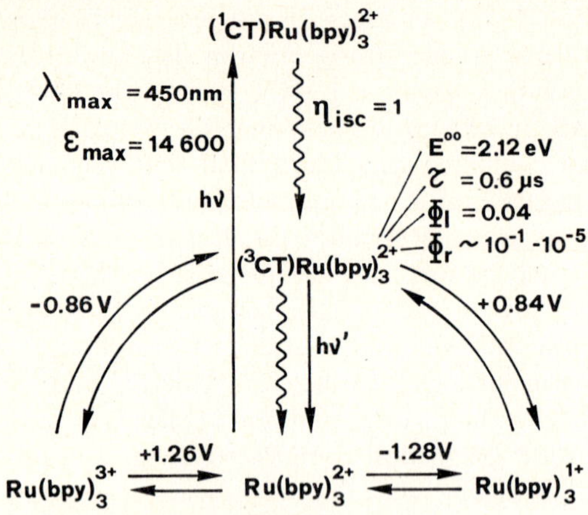

Fig. 12. Schematic representation of some relevant ground- and excited- state properties of $[Ru(bpy)_3]^{2+}$. ^{1}CT and ^{3}CT are the spin-allowed and spin-forbidden metal-to-ligand charge-transfer excited states, responsible for the high intensity absorption band with λ_{max} = 450 nm and the luminescent band with λ_{max} = 615 nm, respectively. The other quantities shown are: intersystem crossing efficiency (η_{isc}); energy (E^{00}) and lifetime (τ) of the ^{3}CT state; luminescence quantum yield (Φ_l); quantum yield for ligand detachment (Φ_r). The reduction potentials of couples involving the ground and the ^{3}CT excited states are also indicated. All the data refer to aqueous solution at room temperature, except Φ_r which refers to a variety of experimental conditions (Sect. 5.4). The potentials are vs NHE

An important aspect concerning the Ru(II)-polypyridine complexes is the possibility to change gradually (i.e., "to tune") the various ground and excited state properties by a judicious choice and combination of the ligands [14, 63]. Recent review articles [45, 68] have shown that more than 200 bidentate polypyridine ligands (L) have been used in the Ru(II) chemistry and that a great number of homoleptic $Ru(L)_3^{2+}$, and bis-heteroleptic $Ru(L)_{3-n}(L')_n^{2+}$ complexes have been prepared, which cover a wide range of values of redox potentials, excited state energies, excited state lifetimes, as shown in the exemplifying sample reported in Table 1. Data concerning a much larger number of Ru(II) complexes are reported in Ref. [45]. Polypyridine Cr(III) [69] and Os(II) [70] complexes also exhibit suitable properties to play the role of **LAS** and **LES**. Finally, it should be pointed out that in the last few years a great number of di- and poly-nuclear complexes of Ru(II), Os(II), and Cr(III), with polypyridine ligands have been investigated. Most of such systems can also be used as **LAS** and **LES** [71].

5.1 Absorption and Emission Spectra

The absorption spectrum of $Ru(bpy)_3^{2+}$ is shown in Fig. 13 along with the proposed assignments [14]. The bands at 185 nm (not shown in the Figure) and 285 nm have been assigned to LC transitions by comparison with the spectrum of protonated bipyridine. The two remaining intense bands at 240 and 450 nm

Table 1. Spectroscopic and electrochemical properties of some selected families of Ru(II)-polypyridine complexes[a]

	Absorption		Emission		E^{00}	Electrochemistry			
	λ_{max} nm	ε M^{-1} cm^{-1}	λ_{max} nm	τ µs	eV	E_{ox} V	E_{red} V	$^*E_{ox}$ V	$^*E_{red}$ V
$Ru(bpy)_3^{2+}$	452	13000	615	1.10	2.13	1.26	−1.35	−0.87	0.78
$Ru(bpy)_2(DM\text{-}bpy)^{2+}$ [b]	448	11500	620	0.74	2.08	1.22	−1.37	−0.86	0.71
$Ru(bpy)(DM\text{-}bpy)_2^{2+}$	453	12200	625	0.72	2.10	1.18	−1.41	−0.92	0.69
$Ru(DM\text{-}bpy)_3^{2+}$	456	10800	625	0.21	2.08	1.15	−1.46	−0.93	0.62
$Ru(bpy)_2(DTB\text{-}bpy)^{2+}$ [b]	450	14500	625	1.17	2.10	1.21	−1.37	−0.89	0.73
$Ru(bpy)(DTB\text{-}bpy)_2^{2+}$	454	15300	630	1.07	2.10	1.16	−1.39	−0.94	0.71
$Ru(DTB\text{-}bpy)_3^{2+}$	456	16800	625	1.15	2.16	1.11	−1.44	−1.05	0.72
$Ru(bpy)_2(biq)^{2+}$	526	8710	742	0.27	1.70	1.33	−0.91	−0.37	0.79
$Ru(bpy)(biq)_2^{2+}$	547	8550	742	0.20	1.69	1.40	−0.82	−0.29	0.87
$Ru(biq)_3^{2+}$	524	9000	—	—	1.73	1.47	−0.73	−0.26	0.83

[a] Acetonitrile solution, room temperature. A much wider compilation is given in ref. [45] where the original sources are also reported;
[b] DM-bpy = 3,3′-dimethyl-bpy, DTB-bpy = 4,4′-di-*tert*-butyl-bpy

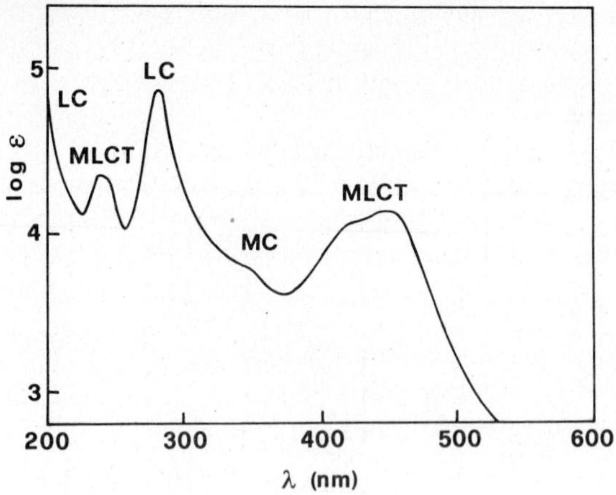

Fig. 13. Absorption spectrum of $Ru(bpy)_3^{2+}$ in aqueous solution at room temperature

have been assigned to MLCT $d \to \pi^*$ transitions. The shoulders at 322 and 344 nm might be due to MC transitions. In the long wavelength tail of the absorption spectrum a shoulder is present at about 550 nm ($\varepsilon \sim 600$) when the absorption measurements are made on $Ru(bpy)_3^{2+}$ in an ethanol-methanol glass at 77 K [72]. This absorption feature is thought to correspond to the lowest MLCT excited state(s).

The displacement of the visible absorption bands to the blue or to the red can readily be obtained by replacing bpy with ligands that are more difficult (e.g. *i*-biq; *i*-biq is 3,3′-biisoquinoline) or easier (e.g. biq; biq is 2,2′-biquinoline) to reduce [45]. For bis-heteroleptic complexes, distinct absorption bands corresponding to the two different types of MLCT transitions may be observed. For tris-

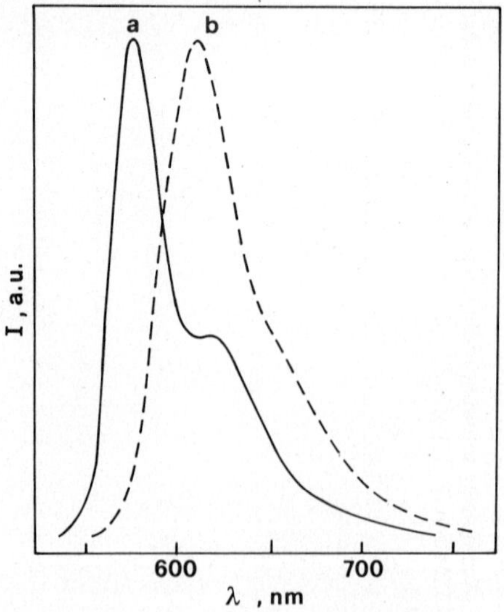

Fig. 14a, b. Luminescence spectrum of $Ru(bpy)_3^{2+}$ in ethanol-methanol solvent: (**a**) 77 K, (**b**) room temperature

heteroleptic complexes [73], the absorption spectrum is still predictable from the spectra of the parent homoleptic complexes, but some degree of interaction between the chromophoric centers based on different ligands can also be expected [74].

The luminescence of $Ru(bpy)_3^{2+}$ (Fig. 14) is a typical ^{3}MLCT emission [14]: (i) it occurs at much lower energy (17200 cm^{-1} at 77 K) than the phosphorescence of free bpy (23100 cm^{-1}); (ii) the luminescence spectrum is structured at low temperature; (iii) the radiative lifetime is ~13 μs. Most of the known Ru(II)-polypyridine complexes exhibit a luminescent behavior quite similar to that of $Ru(bpy)_3^{2+}$, indicating that the lowest excited state is a ^{3}MLCT state. By an appropriate choice of the ligands, however, it has been possible to change the orbital nature of the lowest excited state. ^{3}MC emission can be obtained decreasing the ligand field strength, e.g. replacing one or two polypyridine ligands by Cl^- ions [75]. It should be noted, however, that the σ donor ability of the Cl^- ligand also lowers the energy of the Ru → L MLCT excited states by increasing the negative charge on the metal. Thus, when the $Ru(L)_3^{2+}$ compound has a ^{3}MLCT excited state at very low energy (as is the case for L = biq or DMCH; DMCH is 5,6-dihydro-4,7-dimethyldibenzo [3,2-b:2,3′-j][1,10]phenanthroline), substitution of L by $2Cl^-$ does not cause an inversion on the excited state energy ordering. A case in which a clean ^{3}MC emission can be observed is $Ru(i\text{-}biq)_2Cl_2$ [75]. ^{3}LC emission can be obtained using polypyridine ligands which satisfy the following requirements: (i) presence of a relatively low-lying ^{3}LC level; (ii) sufficiently high ligand field strength to keep ^{3}MC at high energy; (iii) sufficiently negative reduction potential of L to keep ^{3}MLCT at high energy. The *i*-biq ligand apparently satisfies these requirements [76]. A clear example of tuning of the excited state orbital nature is shown in Fig. 15. For $Ru(i\text{-}biq)_3^{2+}$, emission is clearly ^{3}LC in nature, as shown by: (i) the shape of the luminescence spectrum, which is identical to that of the free *i*-biq ligand; (ii) the energy of the emission maximum, which is less than 1000 cm^{-1} red shifted compared with that of the free ligand; (iii) the relatively long emission lifetime

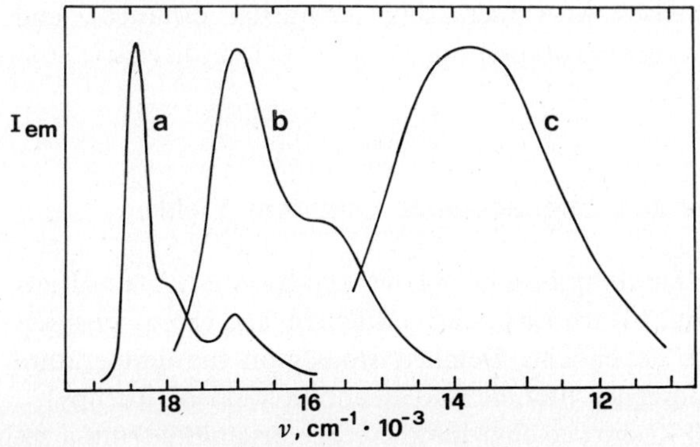

Fig. 15a–c. Luminiscence spectrum of $Ru(i\text{-}biq)_3^{2+}$ (**a**), $Ru(i\text{-}biq)_2(bpy)^{2+}$ (**b**), and $Ru(i\text{-}biq)_2Cl_2$ (**c**) in ethanol-methanol solvent at 77 K [75]

at 77 K (96 μs). When an *i*-biq ligand is replaced by bpy, which has a higher ^{3}LC level, a similar ligand field strength, and it is easier to reduce, the emission (Fig. 15) moves to the red, exhibits a different structure and a shorter lifetime at 77 K (5 μs), maintaining a high intensity in fluid solution at room temperature, as expected for a ^{3}MLCT (specifically Ru → bpy) emitting level [77]. When an *i*-biq ligand is replaced by two Cl^- ligands, the emission moves further to the red, becomes broad, unstructured, shorter lived (2.2 μs), and can no longer be observed at room temperature as expected for a ^{3}MC emission [75].

Interestingly, a tuning between ^{3}LC and ^{3}MLCT can also be obtained by changing the acidity of the solution. This happens for the $Ru(L)_2(CN)_2$ complexes (L = bpy, *i*-biq, phen; phen is 1,10-phenanthroline) [78–80], where the CN^- ligands can be protonated at their nitrogen end. The unprotonated forms of the complexes exhibit the usual ^{3}MLCT emission. Protonation withdraws negative charge from the metal, thus moving the MLCT (Ru → L) levels to higher energy, while the LC levels are unaffected. For the diprotonated species, the emission clearly exhibits ^{3}LC character as shown, for example by the extremely long lifetime (8.8 ms) of $Ru(i\text{-biq})_2(CNH)_2^{2+}$ in H_2SO_4 at 77 K.

Replacement of two bpy ligands of $Ru(bpy)_3^{2+}$ with 4 CN^- ions leads to the anionic $Ru(bpy)(CN)_4^{2-}$ complex whose luminescence is strongly pH and solvent dependent [81, 82].

As mentioned above, for the great majority of Ru(II)-polypyridine complexes the lowest (luminescent) excited state is a ^{3}MLCT level (or a cluster of ^{3}MLCT levels). For systematic studies and practical applications in the field of luminescence, chemiluminescence, and sensitization it is important to have a series of complexes covering a broad range of excited state energies. This can be done by (i) changing the type of ligand involved in the formation of the MLCT excited state, (ii) controlling the amount of negative charge on the metal by changing the nature of the ligands not involved in the excited state, and (iii) changing the solvent. The parameters to be taken into consideration are the reduction potential, the σ-donor ability (which is related to the pK_a), and the π-acceptor properties of the ligands, the charge separation in the excited state (since the stabilizing coulombic interaction between the hole on the metal and the electron on the ligand decreases with increasing separation distance), and solvent parameters which govern the complex-solvent interaction (dielectric constant, donor and acceptor numbers, etc.).

5.2 Excited State Lifetime and Luminescence Quantum Yield

Excitation of Ru(II)-polypyridine complexes in any of their absorption bands leads to a luminescence emission (Fig. 14) whose intensity, lifetime, and energy position are more or less temperature dependent. Detailed studies on the temperature dependence [14] of the luminescence lifetime and quantum yield of $Ru(bpy)_3^{2+}$ in the temperature range 2-70 K showed that luminescence originates from a set of three closely spaced levels in thermal equilibrium. The theoretical description of the luminescent levels has been a matter of debate [49, 83, 84]. There is no doubt about their MLCT nature, and evidence is growing in favor of a localized single-ligand description even in the case of equivalent ligands [85–88].

In rigid alcoholic glass at 77 K the emission lifetime of $Ru(bpy)_3^{2+}$ is ~5 μs and the emission quantum yields is ~0.4 [14]. Taken together with the unitary intersystem crossing efficiency, these figures yield a value of 13 μs for the radiative lifetime. Values of this order of magnitude have been found for MLCT excited states of other transition metal complexes [89–91], while LC excited states exhibit radiative lifetimes in the ms range [62, 89, 90].

With increasing temperature, the emission lifetime (Fig. 16) and quantum yield decrease [45]. To account for this behaviour, $1/\tau$ can be expressed as a sum of a temperature independent and several temperature dependent terms, Eq. (26). The

$$1/\tau = k_0 + \Sigma_i k_i(T) \tag{26}$$

temperature-independent term can be expressed by

$$k_0 = k^r + k_0^{nr} \tag{27}$$

where k^r is the radiative rate constant (usually taken to be temperature independent above 77 K [92]) and k_0^{nr} is a radiationless rate constant related to deactivation to the ground state via a weak-coupling mechanism. The temperature dependent terms can be associated with radiationless processes related, in a schematic way, either to an activated surface crossing to another excited state, described by the Arrhenius equation, Eq. (28), or to the coming into play of effects (e.g.

$$k_i^{nr} = A_i \exp(-\Delta E_i/RT) \tag{28}$$

solvent repolarization) that do not occur at low temperature because of the frozen environment [93–96]. This second type of thermally activated processes

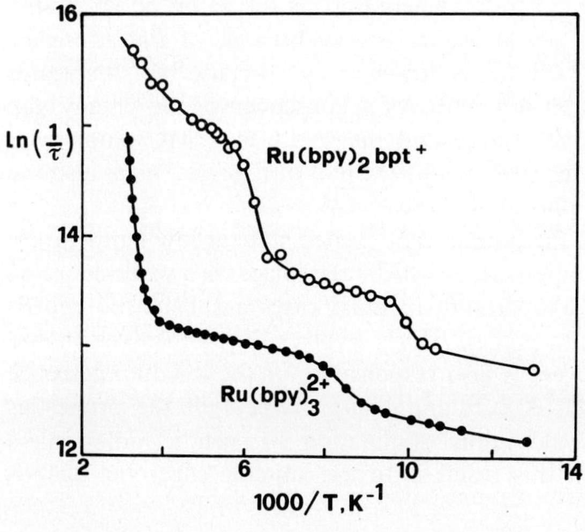

Fig. 16. Temperature dependence of the emission lifetime of $Ru(bpy)_3^{2+}$ and $Ru(bpy)_2bpt^+$ in propionitrile-butyronitrile solution; bpt^- is the anion of 3,5-bis-(pyridin-2-yl)-1,2,4-triazole, see text

can be dealt with by the empirical equation, Eq. (29), which describes a stepwise change of lifetime centered at a certain temperature T_{B_i} [93].

$$k_i^{nr} = \frac{B_i}{1 + \exp\left[C_i(1/T - 1/T_{B_i})\right]} . \qquad (29)$$

In Eq. 29, C_i is a temperature related to the smoothness of the step and B_i is the increment for k_i^{nr} at $T \gg T_{B_i}$. Such an equation is particularly useful for describing the behavior of a system in the glass-fluid region of a solvent matrix.

The radiative rate constant for the various complexes can be obtained from the luminescence lifetime and luminescence quantum yield measured at room temperature:

$$k^r = \Phi_{em}/\tau . \qquad (30)$$

Values of k_0^{nr} can be obtained from the lifetime at 77 K, assuming that the radiative rate constant is temperature independent (vide supra):

$$k_0^{nr}\,(77\ K) = 1/\tau\,(77\ K) - k^r . \qquad (31)$$

The values so obtained are found to increase with decreasing energy of the luminescent level, as expected on the basis of the energy gap law [25, 37].

For a discussion on **LAS** and **LES**, only the room temperature properties are relevant. For $Ru(bpy)_3^{2+}$, at $T > 250$ K the ln $(1/\tau)$ vs $1/T$ plot (Fig. 16) exhibits a steep, linear behavior which is accounted for [77] by an Arrhenius term, Eq. (28), with high frequency factor ($\sim 10^{14}\ s^{-1}$) and large activation energy ($\sim 4000\ cm^{-1}$). This term is associated with an activated surface crossing to an upper lying ^{3}MC excited state which can undergo fast deactivation to ground state and/or ligand dissociation [45, 51, 97, 98] (Sect. 5.4). This behavior is common to most Ru(II)-polypyridine complexes, but noticeable exceptions have also been found. For example, for $Ru(bpy)_2(bpt)^+$ (where bpt^- is the anion of 3,5-bis(pyridin-2-yl)-1,2,4-triazole), ^{3}MC lies at higher energy because of the strong σ-donor power of bpt^-, and ^{3}MLCT lies at lower energy because bpt^- decreases the positive charge on the metal ion [99]. As a consequence, the energy gap between ^{3}MLCT and ^{3}MC is very large and the ^{3}MLCT → ^{3}MC pathway is blocked even at high temperature, as shown by the plot of Fig. 16. This is also the reason why the complex is not photolabile (Sect. 5.4).

In conclusion, the lifetime of the luminescent excited state at low temperature is largely controlled by a radiationless decay which takes place via a weak-coupling mechanism, whereas at room temperature in most cases a substantial contribution comes from a deactivation process which involves thermal activation and strong coupling. The latter process is also responsible for the low luminescence yield and for photodegradation. A substantial improvement of the properties needed for **LAS** and **LES** would imply elimination of such a radiationless decay process. A way to reach this result is to use suitable cage-type ligands (Sect. 5.4).

5.3 Redox Properties

A fundamental requirement to play the roles of **LAS** and **LES** is the stability of the oxidized and/or reduced forms, i.e. the reversibility of the oxidation and/or reduction processes. In the ground state a wealth of information is available on the electrochemical behaviour of ruthenium polypyridine complexes [48–50] as obtained for instance from cyclic voltammetry (CV) in non-aqueous solvents. For many complexes of the Ru(II)-polypyridine family it is possible to localize, with a high degree of certainty, the donor and acceptor orbitals involved in the electrochemical processes.

Oxidation of Ru(II)-polypyridine complexes usually involves a metal centered orbital (t_{2g} in octahedral symmetry), with formation of genuine Ru(III) complexes (low spin $4d^5$ configuration) which are inert to ligand substitution [16]:

$$[Ru^{II}(L)_3]^{2+} \rightarrow [Ru^{III}(L)_3]^{3+} + e^- . \tag{32}$$

Normally, only this metal centered oxidation process is observed in the potential range available, but in SO_2 solution at −70 °C also ligand centered oxidation, at higher potential, has been observed [100]. Comparison of the potentials reported in the literature is not an easy task because of the different experimental conditions and reference electrodes used. It can be stated, however, that the Ru(III)/Ru(II) potentials in most complexes which only contain polypyridines type ligands fall in a rather narrow range around +1.25 V with respect to NHE [45]. Substitution of one or more polypyridine ligands by another coordinating ligand can drastically change these potentials. The substitution of one bpy in $Ru(bpy)_3^{2+}$ with two Cl^- ions to give $Ru(bpy)_2Cl_2$ lowers the potential to about +0.32 V (in AN, vs SCE) [101], whereas the strong π*-acceptor CO causes an increase above +1.9 V in $Ru(bpy)_2(CO)_2^{2+}$ (AN solution, vs SCE) [102].

Reduction of Ru(II)-polypyridine complexes may involve, in principle, either a metal-centered or a ligand-centered orbital, depending on the relative energy ordering. When the ligand field is sufficiently strong and/or the ligands can be easily reduced, reduction takes place on a ligand π* orbital. This is the commonly observed behavior of Ru(II)-polypyridine complexes [48–50, 103]. The reduced form, keeping its low-spin $4d^6$ configuration, is usually inert and the reduction process is reversible:

$$[Ru^{II}(L)_3]^{2+} + e^- \rightarrow [Ru^{II}(L)_2(L^-)]^+ . \tag{33}$$

As mentioned above, the added electron is localized on a single ligand. Several reduction steps can often be observed in the potential range accessible. This range has been considerably enlarged through the measurement of CV in DMF at −54 °C [61, 104]. Up to six electrons can be pumped into $Ru(bpy)_3^{2+}$ in this way [61], yielding a highly reduced complex which can be formulated as $[Ru^{2+}(bpy^{2-})_3]^{4-}$. The localization of the acceptor orbitals in the reduction processes is often particularly clear in mixed-ligand complexes involving polypyridine ligands with different energies of the π* orbitals. In such cases, the first and subsequent reduction steps can be attributed to the various ligands in the complex [45, 105]. This is also the case for the tris-heteroleptic complexes [73]. When the ligand field is weak and/or the ligands cannot be easily reduced, the lowest empty orbital in the reduction process can be metal centered (σ^*_M, e_g parentage in octahedral symmetry). In such a case, reduction leads to an unstable low spin d^7 system, which gives rise to a fast ligand dissociation making the process electrochemically irreversible, Eq. (34).

$$[Ru^{II}(LL)_3]^{2+} + e^- \rightarrow [Ru^I(LL)_3]^+ \rightarrow [Ru^I(LL)_2]^+ + LL \tag{34}$$

It is important to note that, if Koopmans' theorem is valid for the starting t^6_{2g} system and for the one-electron reduced species, the π^*_L or σ^*_M orbitals involved in the reduction processes (redox orbitals) are the same orbitals which are involved in the MLCT and MC transitions, respectively (spectroscopic orbitals) [49]. Thus, reversibility of the first reduction step, indicating a ligand centered LUMO, also implies (to a first approximation) that the lowest excited state is MLCT. More generally, there are important correlations between the electrochemical and spectroscopic data [61, 106–109].

As we have seen in Sect. 2, the excited state redox potentials are given, to a first approximation, by Eqs. 3 and 4, and thus they can be tuned by changing the ground state redox potentials and/or the excited state energy. Since reduction usually takes place on a ligand, the ground state reduction potential will be roughly related to the reduction potential of the free ligand (compare, e.g. the first reduction potentials of $Ru(bpy)_3^{2+}$ and $Ru(biq)_3^{2+}$, Table 1, with those of the free ligands: $E^0(bpy/bpy^-) = -2.22$ V; $E^0(biq/biq^-) = -1.74$ V [105]). However, the ability of a coordinated ligand to accept an electron also depends on the amount of charge transferred to, and received from, the metal by the other ligands via the σ and π bonding. It should also be recalled that the changes in the reduction potential of the ground state do not cause an equal change in the reduction potential of the excited state because the energy of the ^{3}MLCT level decreases as the ligand becomes easier to reduce.

As we have seen above, oxidation of Ru(II)-polypyridine complexes usually consists in the removal of an electron from a $\pi_M(t_{2g})$ metal orbital. The oxidation potential, however, is affected by the nature of the ligands because the amount of electric charge localized on the metal (and thus, its tendency to lose an electron) is governed by the σ-donor and π-acceptor properties of the ligands. For ligands of the same series, the presence of electron withdrawing groups increases the oxidation potential while the opposite occurs, of course, for electron donating substituents. Linear correlations between the reduction potential of the complexes and the Hammet constants of the free ligands have been obtained for 4,4′- and 5,5′-disubstituted bpy derivatives [110]. In conclusion, the factors which govern the excited state redox potentials are known and can be manipulated by changing the nature of the ligands. Hundreds of Ru(II)-polypyridine complexes having variable redox potentials in the excited state (and, of course, in the ground state) are now available [45]. This is particularly useful for systematic studies where homogeneous series of powerful oxidants or reductants are needed.

5.4 Photochemical Stability

$Ru(bpy)_3^{2+}$ is not photochemically inert towards ligand substitution. In aqueous solution the quantum yield of $Ru(bpy)_3^{2+}$ disappearance is in the range 10^{-5} to 10^{-3}, depending on the pH of the solution and temperature. The exact nature of the reaction products has not been elucidated [97, 98]. In low polarity solvents such as CH_2Cl_2 the photochemistry of $[Ru(bpy)_3]X_2$ ($X = Cl^-$, Br^-, NCS^-) is well behaved [98, 111] giving rise to $Ru(bpy)_2X_2$ as final product. The quantum yields are in the range 10^{-1}–10^{-3}. The PF_6^- anion does not give photosubstitution products. A substantial difference between H_2O (dielectric constant $\varepsilon = 80.2$) and CH_2Cl_2 ($\varepsilon = 9.1$) solutions is that salts of $Ru(bpy)_3^{2+}$ are completely ion-paired in the latter medium.

A detailed mechanism for the ligand photosubstitution reaction of $Ru(bpy)_3^{2+}$ in CH_2Cl_2 has been proposed [98] (Fig. 17). According to this mechanism, thermally activated formation of a ^{3}MC excited state (vide supra) leads to the cleavage of a Ru—N bond, with formation of a five-coordinated square pyramidal species. In the absence of coordinating species, as with the PF_6^- salt, this five coordinated intermediate returns to $Ru(bpy)_3^{2+}$. When coordinating solvents or anions are present, a hexacoordinated intermediate containing a monodentate bpy is formed. This species can then undergo loss of bpy and formation of $Ru(bpy)_2X_2$ or a "self-annealing" process (chelate ring closure), with reformation of $Ru(bpy)_3^{2+}$. The "self-annealing" protective step is favored in aqueous solution, presumably because of stabilization of the cationic $Ru(bpy)_3^{2+}$ species, whereas formation of neutral $Ru(bpy)_2X_2$ complexes is favored in low polarity solvents. Photoracemization of $Ru(bpy)_3^{2+}$ [112] also occurs with low quantum yield (2.9×10^{-4} in water at 25 °C). This process can be accounted for by a rearrangement of the square pyramidal primary photoproduct (Fig. 17) into a trigonal bipyramidal intermediate which can lead back to either the Δ or the Λ isomer [98].

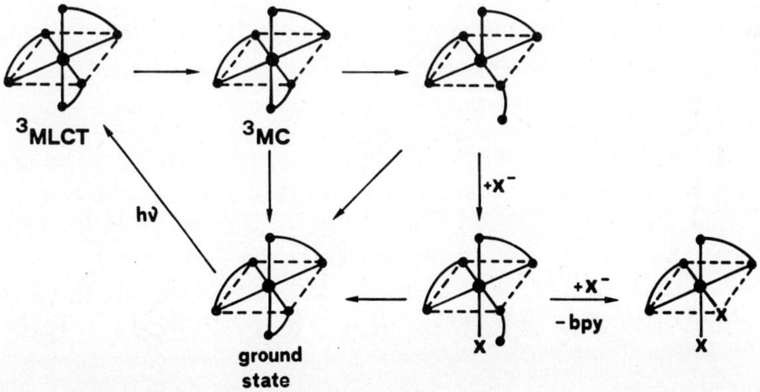

Fig. 17. Schematic representation of the proposed mechanism of ligand photosubstitution reactions for $Ru(bpy)_3^{2+}$ [98]

Ligand photodissociation is, of course, a major drawback for the use of Ru(II)-polypyridine complexes as **LAS** or **LES** (Sect. 3). From the above discussion, it follows that to avoid photolability one should prevent the population of ^{3}MC and/or the ligand dissociation from ^{3}MC. Population of ^{3}MC can be prevented or at least reduced by: (a) addition of sufficient concentration of quencher to capture ^{3}MLCT before surface crossing to ^{3}MC can occur; (b) working at low temperature (Sect. 5.2); (c) increasing the energy gap between ^{3}MLCT and ^{3}MC; (d) increasing pressure [113]. On the other hand, ligand dissociation from ^{3}MC can be reduced by (e) avoiding the use of coordinating anions in solvent of low dielectric constant and (f) linking together the three bpy ligands so as to form a caging ligand structure which encapsulates the metal ion. Points (c) and (f) are particularly interesting for the purpose of light absorption and/or light emission sensitization and much effort is presently devoted to research along such directions.

The tuning of excited state energy by changing ligands (point (c) above) is discussed in Sect. 5.1. Caging Ru^{2+} into a single polypyridine type ligand (point (f)) poses severe synthetic difficulties. Furthermore, subtle problems related to the shape and size of the cage should be carefully considered. If the Ru-N bonds are too long and/or the bite angles are unfavorable, the ligand field strength is small, the lowest ^{3}MC level drops below the lowest ^{3}CT level, and most of the valuable properties of the complex disappear. Molecular models show that this is the case for the bpy.bpy.bpy cryptand shown in Fig. 10b. Such a ligand, in fact, is appropriate for the larger, not symmetry-demanding Eu^{3+} ion, but it is clearly too rigid to allow the octahedral coordination required by Ru^{2+}. This is apparently confirmed by the results obtained in a recent report [114]. The dimension of the cage and the symmetry of the coordination sphere, however, may be "tuned" using spacers of different dimensions (Fig. 18) [115]. Molecular models show that with a larger spacer and three bpy coordinating groups it is possible to obtain a cage-type ligand, more flexible than that shown in Fig. 10b and capable of offering a suitable coordination site for Ru^{2+}. A cage complex

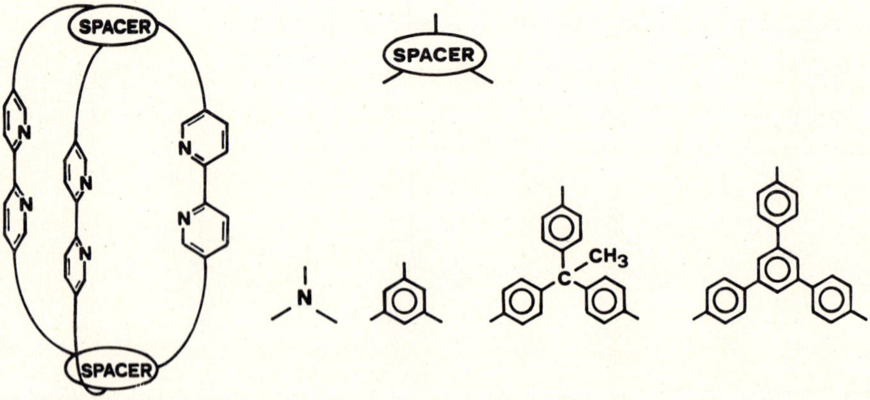

Fig. 18. Design of cages of different dimension for encapsulation of metal ions [115]

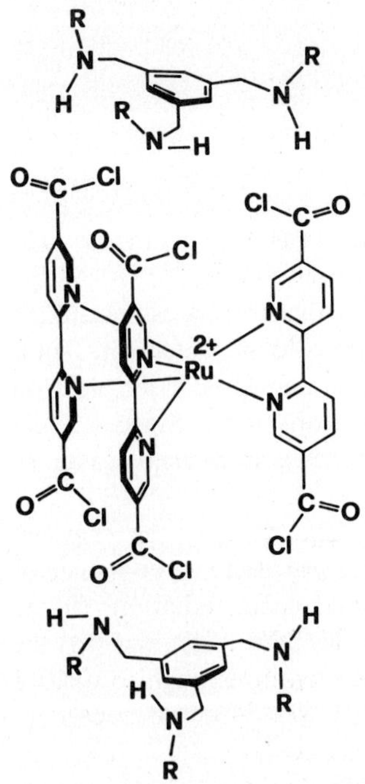

Fig. 19. Design of the template reaction to produce a bpy cage ruthenium complex [116]

was therefore designed and prepared via a template reaction, Fig. 19 [116]. As expected, it shows spectroscopic properties ($\lambda_{max}^{abs} = 455$ nm, $\lambda_{max}^{em} = 612$ nm) quite similar to those of $Ru(bpy)_3^{2+}$, but a longer excited state lifetime at room temperature (1.7 vs 0.8 μs), and a much larger stability toward ligand photosubstitution ($\Phi_r < 10^{-6}$, compared to $\Phi_r = 0.017$ for $Ru(bpy)_3^{2+}$, in CH_2Cl_2 solution containing 0.01 M Cl^-) [117]. This should ensure a quite high turnover number when the bpy-cage Ru complex is used as a **LAS** or **LES.**

6 LAS and LES in Action: Selected Examples

LAS and **LES** can be used for a variety of applications. The examples illustrated in this section underline both the outstanding potentialities and the great difficulties that are encountered in this field.

6.1 Photosensitized Water Splitting

The splitting of H_2O by solar radiation, Eq. (35), is one of the most important

$$H_2O \xrightarrow[\text{radiation}]{\text{solar}} H_2 + 1/2\,O_2 \tag{35}$$

goals of modern chemistry because it would make available a clean and renewable fuel. Water, however, only absorbs in the far UV and therefore it cannot

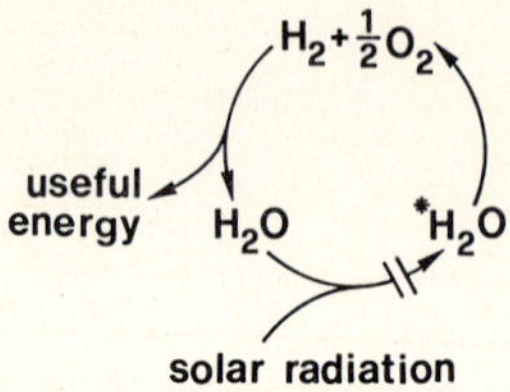

Fig. 20. Direct water splitting by solar energy is prevented by lack of solar light absorption

be electronically excited by solar radiation which is distributed essentially in the visible and IR spectral region. Thus, direct photolysis of water with solar radiation cannot take place (Fig. 20). Reaction (35) is essentially an electron transfer process which involves 1.23 V per electron transferred. Since 1.23 eV corresponds to the energy of 1008 nm photons, Eq. (35), in principle, can be induced by solar radiation via a suitable LAS [118].

Looking at the properties of $Ru(bpy)_3^{2+}$ (Fig. 12), we can see that, in principle, it can play the role of **LAS** in the water splitting process [119] since (i) it is able to absorb a substantial fraction of (visible) solar radiation, (ii) its excited state is a reductant strong enough to reduce H^+, Eq. (36), and (iii) the $Ru(bpy)_3^{3+}$ species produced by Eq. (36) is an oxidant strong enough to oxidize water, being regenerated, Eq. (37).

$$^*Ru(bpy)_3^{2+} + H^+ \rightarrow Ru(bpy)_3^{3+} + 1/2\ H_2 \qquad (36)$$

$$\Delta G^0 = -0.44 \text{ eV at pH 7}$$

$$Ru(bpy)_3^{3+} + 1/2\ H_2O \rightarrow Ru(bpy)_3^{2+} + H^+ + 1/4\ O_2 \qquad (37)$$

$$\Delta G^0 = -0.45 \text{ eV at pH 7}$$

In practice, however, such a cyclic process (Fig. 21) does not work since the reaction between $^*Ru(bpy)_3^{2+}$ and H^+ to yield H_2 is slow because of mechanistic reasons related to its two-electron nature and to the short lifetime of $^*Ru(bpy)_3^{2+}$ (μs time scale). In other words, there is something like a short-circuit which immediately dissipates the absorbed light energy and prevents the involvement of H_2O.

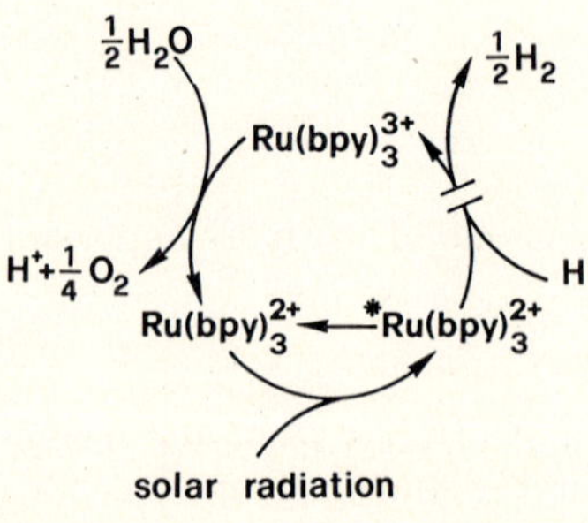

Fig. 21. Scheme of water splitting by solar energy using $Ru(bpy)_3^{2+}$ as photosensitizer. For more details, see text

In order to overcome this difficulty, one should either prolong the excited state lifetime by several orders of magnitude (which is practically impossible) or intercept the excited state before it undergoes deactivation. The latter way is a practical one since it is possible to find species capable of giving a very fast electron transfer reaction with $*Ru(bpy)_3^{2+}$. By a careful choice, it is also possible to find species (relays, Sect. 3.1) that, once reduced by $*Ru(bpy)_3^{2+}$, are capable of reducing H^+ to H_2. One such species is methylviologen, MV^{2+}.

$$*Ru(bpy)_3^{2+} + MV^{2+} \rightarrow Ru(bpy)_3^{3+} + MV^{+} \tag{38}$$

$$k \sim 1 \times 10^9 \ M^{-1} \ s^{-1} \ [120]$$

$$MV^{+} + H^{+} \rightarrow MV^{2+} + 1/2 \ H_2 \ . \tag{39}$$

The resulting system (Fig. 22), however, does not work in practice because the oxidized form of the photosensitizer and the reduced form of the relay undergo a fast back-electron transfer reaction, Eq. (40), which prevents the occurrence of the slower reactions shown by Eqs. (37) and (39).

$$Ru(bpy)_3^{3+} + MV^{+} \rightarrow Ru(bpy)_3^{2+} + MV^{2+} \tag{40}$$

$$k = 4.2 \times 10^9 \ M^{-1} \ s^{-1} \text{ in } H_2O \ , \qquad \mu = 0.16 \ [121]$$

In principle, this difficulty can be overcome speeding up Eqs. (37) and (39) by means of suitable catalysts (e.g. colloidal Pt and RuO_2) (Fig. 23), but this is a very hard task and controversial results have been reported [122, 123]. The problem of getting H_2 and O_2 in separate compartments is another very critical requirement.

Systems have also been studied where the **LAS** is adsorbed on or linked to a solid (electrode, catalyst) so that electrons or holes are photoinjected into the

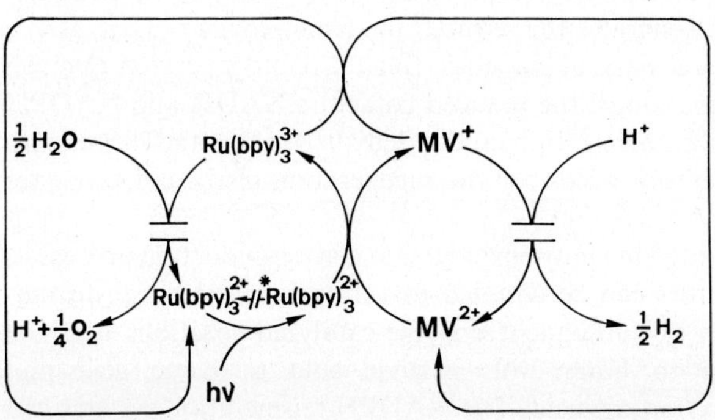

Fig. 22. Scheme of water splitting by a photosensitizer ($Ru(bpy)_3^{2+}$) and a relay (MV^{2+}). For more details, see text

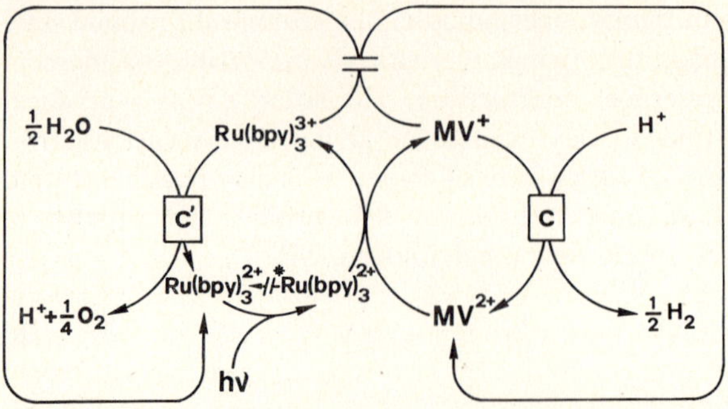

Fig. 23. Scheme of water splitting by a photosensitizer ($Ru(bpy)_3^{2+}$), a relay (MV^{2+}), and two catalysts (colloidal Pt, C, and RuO_2, C') [122, 123]. For more details, see text

substrate (see, e.g. [124, 125]). These and other heterogeneous systems [126] will not be discussed for space reason. For the same reason, sacrificial systems [127] and recent studies dealing with photoinduced charge separation in organized supramolecular structure [6, 71, 128] are not discussed.

6.2 Synthetic Applications of LAS

A **LAS** can be used to carry on reactions of synthetic interest, but this field has not yet been fully explored. A few early examples in the field of organic chemistry [129, 130] have been recently followed by systematic studies on artificial photosystems that use enzymes as biocatalysts in redox reactions that follow the primary photochemical process [131–135]. Such processes are initiated by the absorption of visible light ($\lambda > 400$ nm) by the photosensitizer, yielding an electron-transfer product, A^-, which can serve as a reductant. Provided that this product is recognized by the appropriate enzyme, it can mediate the photoinduced synthesis of useful materials or fuels.

An example of such an artificial photosystem is given by $Ru(bpy)_3^{2+}$ as a photosensitizer, MV^{2+} as an electron acceptor, and $(NH_4)_3EDTA$ as sacrificial electron donor. This system generates the reduced methyl viologen radical, MV^+, which is recognized by several natural enzymes. Such a photogenerated electron carrier can mediate the formation of the reduced cofactors NADH and NADPH in the presence of the biocatalysts lipoamide dehydrogenase and ferredoxin reductase, respectively. Quantum yields for the regeneration of the cofactors are in the range of 0.5–8%.

As these cofactors are essential in many enzymatic oxidation-reduction processes, the photoregenerated cofactors can be coupled to various synthetic transformations. Some examples [132] of subsequent enzyme-catalyzed reactions are: the reduction of butane-2-one to butan-2-ol, pyruvic acid to lactic acid and acetoacetic acid to β-hydroxybutyric acid. The NADPH regeneration systems can also be applied in CO_2-fixation processes [133, 134]. Similarly, the enzymes nitrate reductase and nitrite reductase can be coupled to the artificial photogenerated electron carrier MV^+, thus allowing the effective photosensitized reduction of nitrate (NO_3^-) to ammonia (NH_3) [135].

6.3 Photogalvanic Effect

The conversion of light energy into electrical energy (via intermediate conversion to chemical energy) is also possible using $Ru(bpy)_3^{2+}$ as **LAS**. Consider, for example, a cell consisting of two identical compartments separated by a sintered glass disk and containing a Pt electrode and an aqueous solution of $Ru(bpy)_3^{2+}$ and Fe^{3+} [136]. If one compartment is illuminated and the other is kept in the dark, an electrical potential is generated which depends on the incident light intensity (photogalvanic effect [137]). This potential is due to the difference in the composition of the solutions in the dark and illuminated compartments caused by the following reactions (Fig. 24):

$$Ru(bpy)_3^{2+} + h\nu \rightarrow {}^*Ru(bpy)_3^{2+} \tag{41}$$

$$^*Ru(bpy)_3^{2+} + Fe^{3+} \rightarrow Ru(bpy)_3^{3+} + Fe^{2+} \, . \tag{42}$$

6.4 Electrochemiluminescence

The involvement of **LES** in electrochemical processes may allow the conversion of electrical energy into light. This phenomenon, which is the reverse of the previously described photogalvanic effect, is called electrochemiluminescence (ECL). One of the most noticeable examples of ECL is that concerning $Ru(bpy)_3^{2+}$ in acetonitrile solution [16, 138]. When cyclic square waves between the potentials of formation of $Ru(bpy)_3^+$ and $Ru(bpy)_3^{3+}$ (Fig. 12) are applied at a Pt electrode immersed in the solution, a beautiful luminescence is observed which continues

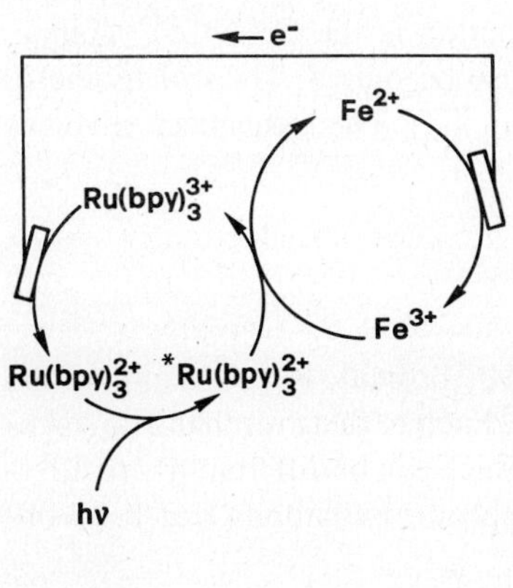

Fig. 24. An example of the use of $Ru(bpy)_3^{2+}$ as LAS: light energy is converted into electrical energy (photogalvanic effect) [136]

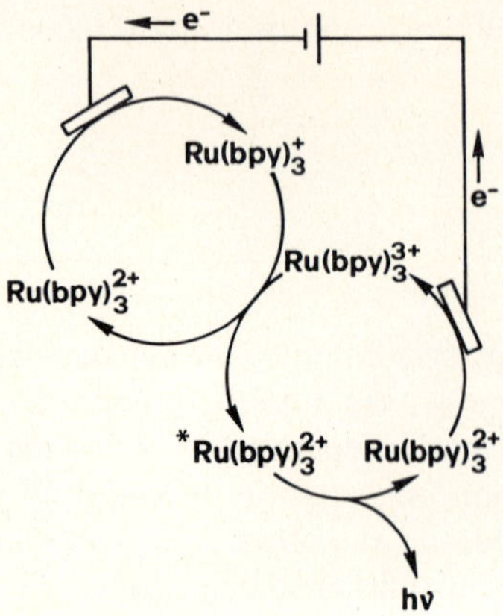

Fig. 25. An example of the use of $Ru(bpy)_3^{2+}$ as LES: electrical energy is converted into light energy [138]

indefinitely if the electrical potential is maintained. The reaction mechanism, illustrated in Fig. 25, involves the following reactions:

$$Ru(bpy)_3^{2+} + e^- \rightarrow Ru(bpy)_3^+ \tag{43}$$

$$Ru(bpy)_3^{2+} - e^- \rightarrow Ru(bpy)_3^{3+} \tag{44}$$

$$Ru(bpy)_3^+ + Ru(bpy)_3^{3+} \rightarrow {}^*Ru(bpy)_3^{2+} + Ru(bpy)_3^{2+} \tag{45}$$

$${}^*Ru(bpy)_3^{2+} \rightarrow Ru(bpy)_3^{2+} + h\nu . \tag{46}$$

6.5 Oscillating Chemiluminescence

The most curious process in which $Ru(bpy)_3^{2+}$ plays the role of **LES** is an oscillating reaction. It is well established that certain types of chemical reactions, under appropriate experimental conditions, organize themselves spontaneously to give rise to regular spacial patterns or to periodic rate fluctuations (see, e.g. [139]). The best studied among the oscillating homogeneous processes is the classical Belousov-Zhabotinskii (BZ) reaction [140], in which a crucial role is played by a redox catalyst. The usual catalyst of the BZ reaction is the Ce^{4+}/Ce^{3+} couple, but polypyridine complexes of Fe and Ru have also been used. The stoichiometry of the overall process is thought to be that of Eq. (47). The mechanism involves

$$2\,BrO_3^- + 3\,CH_2(COOH)_2 + 2\,H^+ \xrightarrow{ox/red} 2\,BrCH(COOH)_2 + 3\,CO_2 + 4\,H_2O \tag{47}$$

oxidation of the reduced form of the catalyst (red) by bromate and reduction of the oxidized form of the catalyst (ox) by malonic acid. Each of these mechanistic stages is complicated and involves many steps, some of which can be different when different catalysts are used. With $Ru(bpy)_3^{2+}$ as a catalyst, oscillations had been observed in the rate of heat evolution [141], in the rate of accumulation of monobromomalonic acid [141], and in the concentration of $Ru(bpy)_3^{3+}$ [142]. From previous experiments [44] it was also known that $Ru(bpy)_3^{3+}$ can be reduced by carboxylic acids with formation of $^*Ru(bpy)_3^{2+}$ and consequent light emission. Coupling these two different pieces of information, it was thought that the BZ oscillating reaction must produce an oscillating chemiluminescence. When the reaction was carried out in a spectrofluorimeter, an oscillating signal was indeed recorded although the light emission was too weak to be observed by eye [143]. This chemical system can be considered as an artificial firefly.

6.6 Lyoluminescence

Dissolution of γ-irradiated NaCl in aqueous solution containing $Ru(bpy)_3^{2+}$ causes the characteristic $^*Ru(bpy)_3^{2+}$ luminescence [144]. The reaction mechanism of this process, where $Ru(bpy)_3^{2+}$ mediates the conversion into light of chemical energy stored in γ-irradiated solids (sensitized lyoluminescence), involves a complicated sequence of redox processes schematized in Fig. 26. Upon dissolution of

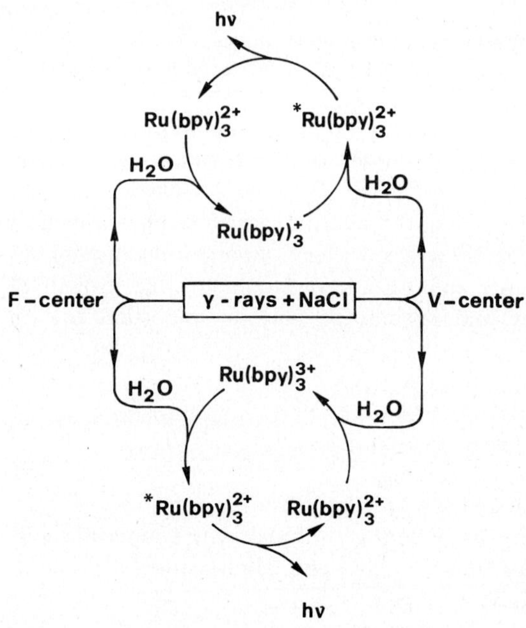

Fig. 26. An example of the use of $Ru(bpy)_3^{2+}$ as LES: chemical energy stored into γ-irradiated solids is converted into light (sensitized lyoluminescence) [144]

γ-irradiated NaCl in water, hydrated electrons and hydrated chlorine radicals are formed from the F- and V-centers present in the crystal:

$$\text{F-center} \xrightarrow{H_2O} e_{aq}^- \tag{48}$$

$$\text{V-center} \xrightarrow{H_2O} Cl_{aq} \tag{49}$$

These species or other reductants and oxidants obtained via secondary reactions cause the formation of $Ru(bpy)_3^+$ and $Ru(bpy)_3^{3+}$. The comproportionation reaction between $Ru(bpy)_3^+$ and $Ru(bpy)_3^{3+}$, the reaction of $Ru(bpy)_3^+$ with a strong oxidant, and/or the reaction of $Ru(bpy)_3^{3+}$ with a strong reductant can produce $^*Ru(bpy)_3^{2+}$ which gives luminescence.

7 Conclusions

Several important electron transfer reactions can be induced by light or can produce light only when they are "mediated" by suitable molecules that can be called light absorption sensitizers **(LAS)** or light emission sensitizers **(LES).** To play this precious role of mediators, **LAS** and **LES** must satisfy several severe requirements. Some families of transition metal complexes, particularly the Ru(II)-polypyridine one, exhibit spectroscopic, photophysical and electrochemical properties that can be tuned so as to meet most of the requirements needed. Recent investigations have also led to the design and synthesis of cage-type Ru(II)-polypyridine complexes which should guarantee the photochemical stability needed for high turnover numbers of **LAS** and **LES** in processes of applicative interest.

All the presently known **LAS** work on the one photon/one electron basis and thus must be helped by electron storing devices (e.g. heterogeneous catalysts) to mediate photoinduced multielectron redox processes (e.g. water splitting). Current investigations concerning the development of luminescent and redox-reactive polynuclear transition metal complexes [71] might lead to **LAS** capable to overcome this drawback.

Apart from their possible use for practical applications such as photochemical conversion of solar energy and chemi-(or electrochemi-)luminescent display devices, the processes based on **LAS** and **LES** have stimulated the growth of important branches of chemistry and promise to play an important role in several future developments.

8 Acknowledgments

We wish to thank Mr. G. Gubellini and Mr. L. Ventura for the drawings.
Financial support for our research from the Progetto Finalizzato Chimica Fine II of the Italian National Research Council (CNR) and Ministero della Pubblica Istruzione is gratefully acknowledged.

9 References

1. Conolly JS (ed) (1981) Photochemical conversion and storage of solar energy, Academic, New York
2. Rabani J (ed) (1982) Photochemical conversion and storage of solar energy, Weizman, Israel

3. Graetzel M (ed) (1983) Energy resources through photochemistry and catalysis, Academic, New York
4. Special volume (1985) J Photochem 29
5. Special issue (1987) New J Chem 11: 77–207
6. Norris JR Jr, Meisel D (eds) (1989) Photochemical energy conversion, Elsevier, New York
7. Adamson AW (1983) J Chem Ed 60: 797
8. Balzani V, Bolletta F, Gandolfi MT, Maestri M (1978) Top Curr Chem 75: 1
9. Balzani V, Bolletta F, Ciano M, Maestri M (1983) J Chem Ed 60: 447
10. Marcus RA, Sutin N (1985) Biochim Biophys Acta 811: 265
11. Balzani V, Bolletta F (1981) J Photochem 17: 479
12. Birks JB (1970) Photophysics of aromatic molecules, Wiley, New York
13. Turro NJ (1978) Modern molecular photochemistry, The Benjamin/Cummings Publishing Co Inc, Menlo Park
14. Crosby GA (1975) Acc Chem Res 8: 231
15. Freed DJ, Faulkner LR (1971) J Am Chem Soc 93: 2097
16. Tokel-Takvoryan NE, Hemingway RE, Bard AJ (1973) J Am Chem Soc 95: 6582
17. Marcus RA (1964) Annu Rev Phys Chem 15: 155
18. Hush NS (1968) Electrochim Acta 13: 1005
19. Sutin N (1983) Progr Inorg Chem 30: 441
20. Meyer TJ (1987) In: Wilkinson G, Gillard RD, McCleverty JA (eds) Comprehensive coordination chemistry, Pergamon, Oxford p 331
21. Wasielewski MR (1988) In: Fox MA, Chanon M (eds) Photoinduced electron transfer Part A, Elsevier, Amsterdam, p 161
22. Efrima S, Bixon M (1976) Chem Phys 63: 4358
23. Kestner NR, Logan J, Jortner J (1974) J Phys Chem 78: 2148
24. Ulstrup J (1979) Charge transfer in condensed media, Springer, Berlin Heidelberg New York
25. Englman R, Jortner J (1970) J Mol Phys 18: 145
26. Marcus RA, Siders P (1982) J Phys Chem 86: 662
27. Indelli MT, Ballardini R, Scandola F (1984) J Phys Chem 88: 2547
28. Rehm D, Weller D (1970) Isr J Chem 8: 259
29. Balzani V, Scandola F (1983) In: Graetzel M (ed) Energy resources through photochemistry and catalysis, Academic, London, p 1
30. Miller JR, Beitz JV, Huddleston RK (1984) J Am Chem Soc 106: 5057
31. Gould IR, Moser JE, Armitage B, Farid S, Goodman JL, Herman MS (1989), J Am Chem Soc 111: 1917
32. Ohno T, Yoshimura A, Shioyama H, Mataga N (1987) J Phys Chem 91: 4365
33. Closs LG, Miller JR (1988) Science 240: 440
34. Connolly JS, Bolton JR (1988) In: Fox MA, Chanon M (eds) Photoinduced electron transfer, Part D, Elsevier, Amsterdam, p 303
35. Marcus RA (1989) J Phys Chem 93: 3078
36. Meyer TJ, Caspar JV (1983) Inorg Chem 22: 2444
37. Meyer TJ (1986) Pure Appl Chem 58: 1193
38. McGuire M, McLendon G (1986) J Phys Chem 90: 2549
39. Rips I, Klafter J, Jortner J, in Ref 6, p 1
40. Closs GL, Piotrowiak P, Miller JR, in Ref 6, p 23
41. Penfield KW, Miller JR, Paddon-Row MN, Cotsaris E, Oliver AM, Hush NS (1987) J Am Chem Soc 109: 5061
42. Barber JR Jr, Gould ES (1971) J Am Chem Soc 93: 4045
43. Young RC, Meyer TJ, Whitten DG (1975) J Am Chem Soc 97: 4781
44. Rubinstein I, Bard AJ (1981) J Am Chem Soc 103: 512
45. Juris A, Balzani V, Barigelletti F, Campagna S, Belser P, von Zelewsky A (1988) Coord Chem Rev 84: 85
46. Balzani V, Carassiti V (1970) Photochemistry of coordination compounds, Academic, London
47. Lever ABP (1984) Inorganic electronic spectroscopy, 2nd edn, Elsevier, Amsterdam
48. Vlcek AA (1982) Coord Chem Rev 43: 39
49. DeArmond MK, Carlin CM (1981) Coord Chem Rev 36: 325

50. Gosh BK, Chakravorty A (1989) Coord Chem Rev 95: 239
51. Kemp TJ (1980) Progr Reaction Kinetics 10: 301
52. Basolo F, Pearson RG (1967) Mechanisms of inorganic reactions, Wiley, New York
53. Moellern T (1972) Lanthanides and actinides In: Emeleus HJ, Bagnall KW (eds) MTP international review of science: Inorganic chemistry, series one, University Park Press, Baltimore, p 289
54. Lehn JM, Sauvage JP (1975) J Am Chem Soc 97: 6700
55. Rodriguez-Ubis JC, Alpha B, Plancherel D, Lehn JM (1984) Helv. Chim Acta 67: 2264
56. Balzani V (1989) Gazz Chim Ital 119: 311
57. Sargeson (1984) Pure Appl Chem 56: 1603
58. Lehn JM (1988) Angew Chem Int Ed Engl 27: 89
59. Lilie J, Shinoara N, Simic MG (1976) J Am Chem Soc 98: 6516
60. Creaser II, Gene RJ, Harrowfield JM, Herlt AJ, Sargeson AM, Snow MR, Springborg J (1982) J Am Chem Soc 104: 6016
61. Ohsawa Y, Hanck KW, De Armond MK (1984) J Electroanal Chem 175: 229
62. De Armond MK, Hillis JE (1971) J Chem Phys 54: 2247
63. Crosby GA, Watts RJ, Carstens DHW (1970) Science 170: 1195
64. De Armond MK (1974) Acc Chem Res 7: 309
65. Kalyanasundaram K (1982) Coord Chem Rev 46: 159
66. Watts RJ (1983) J Chem Ed 60: 835
67. Seddon EA, Seddon KR (1984) The chemistry of ruthenium, Elsevier, Amsterdam
68. Krause RA (1987) Structure and Bonding 67: 1
69. Jamieson MA, Serpone N, Hoffman MZ (1981) Coord Chem Rev 39: 121
70. Kober EM, Caspar JV, Sullivan BP, Meyer TJ (1988) Inorg Chem 27: 4587
71. Scandola F, Chiorboli C, Indelli MT, Chapter 3 in this book
72. Demas JN, Crosby GA (1971) J Phys Chem 79: 991
73. Juris A, Campagna S, Balzani V, Gremaud G, von Zelewsky A (1988) Inorg Chem 27: 3652
74. Myrick ML, Blakley RL, De Armond MK (1989) J Phys Chem 93: 3936
75. Belser P, von Zelewsky A, Juris A, Barigelletti F, Balzani V (1983) Gazz Chim Ital 113: 731
76. Belser P, von Zelewsky A, Juris A, Barigelletti F, Tucci A, Balzani V (1982) Chem Phys Lett 89: 101
77. Barigelletti F, Juris A, Balzani V, Belser P, von Zelewsky A (1983) Inorg Chem 22: 3335
78. Peterson SH, Demas JN (1986) J Am Chem Soc 98: 7880
79. Juris A, Barigelletti F, Balzani V, Belser P, von Zelewsky A (1987) J Chem Soc Faraday Trans 2 83: 2295
80. Davila J, Bignozzi CA, Scandola F (1989) J Phys Chem 93: 1373
81. Scandola F, Indelli MT (1988) Pure & App Chem 60: 973
82. Kato M, Yamauchi S, Hirota N (1989) J Phys Chem 93: 3422
83. Myrick ML, Blakley RL, DeArmond MK, Arthur ML (1988) J Am Chem Soc 110: 1325
85. Krausz, E (1988) Inorg Chem 27: 2392
85. Carrol PJ, Brus LE (1987) J Am Chem Soc 109: 7613
86. Yabe T, Anderson DR, Orman LK, Chang YJ, Hopkins JB (1989) J Phys Chem 93: 2302
87. Bradley PG, Kress N, Hornberger BA, Dallinger RF, Woodruff WH (1981) J Am Chem Soc 103: 7441
88. Forster M, Hester RE (1981) Chem Phys Letters 81: 42
89. Sprouse S, King KA, Spellane PJ, Watts RJ (1984) J Am Chem Soc 106: 6647
90. King KA, Spellane PJ, Watts RJ (1985) J Am Chem Soc 107: 1431
91. Maestri M, Sandrini D, Balzani V, Chassot L, Jolliet P, von Zelewsky A (1985) Chem Phys Lett 122: 375
92. Caspar JV, Meyer TJ (1983) Inorg Chem 22: 2444
93. Barigelletti F, Juris A, Balzani V, Belser P, von Zelewsky A (1987) J Phys Chem 91: 1095
94. Kitamura N, Sato M, Kim HB, Obata R, Tazuke S (1988) Inorg Chem 27: 651
95. Danielson E, Lumpkin RS, Meyer TJ (1987) J Phys Chem 91: 1305
96. Hiraga T, Kitamura N, Kim HB, Tazuke S, Mori N (1989) J Phys Chem 93: 2940
97. Van Houten J, Watts RJ (1976) J Am Chem Soc 98: 4853

98. Durham B, Caspar JV, Nagle JK, Meyer TJ (1982) J Am Chem Soc 104: 4803
99. Barigelletti F, De Cola L, Balzani V, Hage R, Haasnoot JG, Reedijk J, Vos JG (1989) Inorg Chem 28: 4344
100. Garcia E., Kwak J, Bard A (1988) Inorg Chem 27: 4377
101. Sullivan BP, Conrad D, Meyer TJ (1985) Inorg Chem 24: 3640
102. Pinnick DV, Durham B (1984) Inorg Chem 23: 1440
103. Saji T, Aoyagui S (1975) J Electroanal Chem Interfacial Electrochem 58: 401
104. Ohsawa Y, Whangho MH, Hanck KW, De Armond MK (1984) Inorg Chem 23: 3426
105. Belser P, von Zelewsky A (1980) Helv Chim Acta 63: 1675
106. Dodsworth ES, Lever ABP (1986) Chem Phys Lett 124: 152
107. Kawanishi Y, Kitamura N, Kim Y, Tazuke S (1984) Riken Q 78: 212
108. Barigelletti F, Juris A, Balzani V, Belser P, von Zelewsky A (1987) Inorg Chem 26, 4115
109. Ernst SD, Kaim W (1989) Inorg Chem 28: 1520
110. Skarda V, Cook MJ, Lewis AP, McAuliffe GSG, Thomson AJ, Robbins DJ (1984) J Chem. Soc, Perkin Trans 2: 1309
111. Gleria M, Minto F, Beggiato G, Bortolus P (1978) J Chem Soc Chem Commun 285
112. Porter GB, Sparks RH (1980) J Photochem 13: 123
113. Fetterolf ML, Offen HW (1986) J Phys Chem 90: 1828
114. Duerr H, Zengerle K, Trierweiler HP (1988) Z Naturforsch Teil B 43: 361
115. Grammenudi S, Voegtle F (1986) Angew Chem Int Ed Engl 25: 1122
116. Belser P, De Cola L, von Zelewsky A (1988) J Chem Soc Chem Commun 1057
117. De Cola L, Barigelletti F, Balzani V, Belser P, von Zelewsky A, Voegtle F, Ebmeyer F, Grammenudi S (1988) J Am Chem Soc 110: 7210; Barigelletti F, De Cola L, Balzani V, Belser P, von Zelewsky A, Voegtle F, Ebmeyer F (1989) J Am Chem Soc 111: 4662
118. Balzani V, Moggi L, Manfrin MF, Bolletta F, Gleria M (1975) Science 189: 852
119. Creutz C, Sutin N (1975) Proc Natl Acad Sci USA 72: 2858
120. Hoffman MZ, Bolletta F, Moggi L, Hug GL (1989) J Phys Chem Ref Data 18: 219
121. Chiorboli C, Indelli MT, Rampi Scandola MA, Scandola F (1988) J Phys Chem 92: 156
122. Kalyanasundaram K Graetzel M (1979) Angew Chem Int Ed Engl 18: 701
123. Harriman A (1984) J Photochem 25: 33
124. Wrighton MS (1985) Comments Inorg Chem 4: 269
125. Graetzel M (1988) in: (Eds Fox MA, Chanon M), Photoinduced electron transfer part D Elsevier, Amsterdam p 394
126. Kalyanasundarum K (1987) Photochemistry in microheterogeneous systems, Academic Press, London
127. Serpone N (1988) In: Fox MA, Chanon M (eds) Photoinduced electron transfer Part D, Elsevier, Amsterdam p 47
128. Meyer TJ (1989) Acc Chem Res 22: 163
129. Moggi L, Juris A, Sandrini D, Manfrin MF (1984) Rev Chem Interm 5: 107
130. Takagi K, Aoshima K, Sawaki Y, Iwamura H (1985) J Am Chem Soc 107: 47
131. Mandler D, Willner I (1984) J Am Chem Soc 102: 5352
132. Mandler D, Willner I (1986) J Chem Soc Perkin Trans II 805
133. Willner I, Mandler D, Riklin A (1986) J Chem Soc Chem Commun 1022
134. Mandler D, Willner I (1988) J Chem Soc Perkin Trans II 997
135. Lapidot N, Riklin A, Willner I (1989) J Am Chem Soc 111: 1883
136. Lin CT, Sutin N (1976) J Phys Chem 80: 97
137. Lichtin NN (1977) In: Bolton JR (ed) Solar power and fules, Academic, New York, chap 5
138. Wallace WL, Bard AJ (1979) J Phys Chem 83: 1350
139. Noyes RM, Field RJ (1977) Acc Chem Res 10: 273
140. Zhabotinskii AM (1980) Ber Bunseges Phys Chem 84: 303
141. Koros E, Burger M, Friedrich V, Ladanyi L, Nagy Zs, Orban M (1974) Faraday Symp. Chem Soc 9: 28
142. Demas NJ, Diemente D (1973) J Chem Ed 50: 357
143. Bolletta F, Balzani V (1982) J Am Chem Soc 104: 4250
144. Bolletta F, Mulazzani QG, Venturi N, Serpone N, Balzani V (1985) Gazz Chim Ital 115: 137

Photoinduced Electron and Energy Transfer in Polynuclear Complexes

Franco Scandola, Maria Teresa Indelli, Claudio Chiorboli, and Carlo Alberto Bignozzi

Dipartimento di Chimica dell'Università, Centro di Fotochimica CNR, 44100 Ferrara, Italy

Table of Contents

The photochemistry and photophysics of polynuclear transition metal complexes is a new and rapidly developing area of inorganic photochemistry. The photophysical and photochemical behavior of these multi-component systems is characterized by the widespread occurrence of intramolecular, intercomponent electron and energy transfer processes. The aim of this review article is to provide a general overview of the field as it has been developing during the last ten years. The article includes a general introduction with some background material, a detailed survey of the literature, and some projections towards the future of this research field.

1 Introduction

The first pioneering studies on the photochemical behavior of coordination compounds date back to the late 1950s [1–4]. In these three decades, the field has experienced an anormous growth, and the photochemistry of coordination compounds stands now as a well established and active research field [5–13]. The first period of this development (extending to the late 1970s) saw strong activity in the qualitative and quantitative characterization of the photoreactivity of several classes of coordination compounds, notable examples being Cr(III) and Co(III) complexes. Parallel efforts were devoted, in the same period, to the characterization of the photophysical behavior of several classes of coordination compounds, notably Cr(III), Ru(II), Os(II), Rh(III) and Ir(III) complexes (this topic is still being investigated actively, stimulated by the continuous progress in experimental methods). An important cornerstone in the field was the recognition [14], in the early 1970s, that excited coordination compounds can be easily involved in *bimolecular* processes such as energy and electron transfer. In particular in this last process, coordination compounds are extremely versatile because of their easily tunable redox properties. This has led to an extraordinary blossoming of studies on bimolecular quenching processes of excited transition metal complexes [15–20]. In these bimolecular quenching studies, $Ru(bpy)_3^{2+}$, one of the most popular single molecules of the whole chemical scene over the past twenty years [21–24], has certainly played the leading role. The problem of chemical conversion of solar energy [25–29] has provided additional driving force to the field. At the end of this three-decade period, it can be fairly stated that for several classes of simple coordination compounds a satisfactory degree of understanding of the excited-state properties (photophysical behavior, unimolecular reactivity, and bimolecular processes) has been reached.

With this background firmly established, inorganic photochemists have recently started to shift towards the study of more complex systems. This tendency (that is part of a more general trend of today's chemical research from molecules towards supramolecular systems [30–32]) has brought into the photochemical scene *polynuclear* coordination compounds, i.e. relatively large molecules containing two or more transition metal complex subunits ("components") linked together by suitable bridging ligands. The photochemistry and photophysics of polynuclear complexes are expected and found to be more complex (and more intriguing) than those of their simple mononuclear analogues for two main reasons: (i) the photochemical and photophysical properties of each metal-containing component subunit may undergo perturbations upon incorporation into the polynuclear system; (ii) a number of new processes involving different metal-containing subunits (intercomponent processes) may take place in the polynuclear complex, in addition to those (however perturbed) that characterize the behavior of the individual subunits (intra-component processes). Important inter-component photochemical and photophysical processes in polynuclear complexes are, of course, *electron* and *energy transfer*. The study of such processes in polynuclear complexes offers conceptual advantages (absence of diffusional effects, well-defined geometry) and additional mechanistic issues (distance and orientational effects, role of the bridging ligands) with respect to that of analogous bimolecular reactions. Moreover, interesting developments of this field can be imagined if the spatial organization inherent to polynuclear complexes is coupled to the possibility of performing appropriate sequences of intercomponent transfer processes. It is indeed conceivable that research in this area can lead in the future to the development of polynuclear systems capable of performing valuable light-induced functions such as charge separation and vectorial transport of electronic energy ("photonic molecular devices") [33, 34].

The aim of this article is to provide an overview of the photochemistry and photophysics of polynuclear complexes. Whereas in a broad sense all species containing two or more transition metal centers could be included into such a class of compounds, we will adopt a more restrictive definition and limit our discussion to compounds (i) of finite dimensions on a molecular scale and (ii) in which the connections between the metal centers are provided by bridging ligands. Thus, a number of photochemically interesting systems such as polymer-bound metal complexes [35], protein-bound metal complexes [36], or compounds with direct metal-metal bonds [37], will not be dealt with here. Also, systems that contain covalently bound metal porphyrin subunits [38–41] will not be dealt with in this article, in view of the ancillary role played by the metal ion in this effectively organic chromophore. To facilitate the discussion, some background material is briefly presented in Sect. 2. Section 3 contains a survey of recent work on the photochemistry and photophysics of polynuclear transition metal complexes, with particular emphasis on the inter-component energy and electron transfer processes occurring in these systems. Some prospects for the future of this research area, bearing particularly on the concept of photonic molecular devices, are given in Sect. 4.

2 General Background

In this section, some background material that could help in the subsequent discussion (Sect. 3) of the photochemical and photophysical behavior of polynuclear transition metal complexes is briefly presented. This material includes kinetic concepts on electron transfer processes (Sect. 2.1), optical and photoinduced electron transfer (Sect. 2.2), electronic energy transfer processes (Sect. 2.3), and a discussion of the concept of localization as applied to polynuclear complexes (Sect. 2.4).

2.1 Kinetics of Electron Transfer Processes

Electron transfer is a weak-interaction chemical process in which no bond breaking or forming is involved. This permits a simple description of the whole reaction coordinate in terms of the known properties of reactants and products. The classic model of electron transfer reactions is that developed by Marcus, Hush, and Sutin [42–46], (hereafter referred to as the Marcus model). Although more elaborated quantum mechanical models of electron transfer processes have been subsequently developed [47], the Marcus model is still very widely used, as it combines a relatively simple formalism with a remarkable amount of physical insight and predictive power.

According to this model, the rate constant of a unimolecular electron transfer process

$$A.B \xrightarrow{k} A^+.B^- \tag{1}$$

can be expressed in a way reminiscent of conventional transition-state theory, as [45]

$$k = \nu_N k_{el} \exp(-\Delta G^{\neq}/RT) \tag{2}$$

The meaning of the various terms in Eq. 2 can be conveniently discussed in terms of the energy profiles of Fig. 1. In this figure, the two curves represent the potential

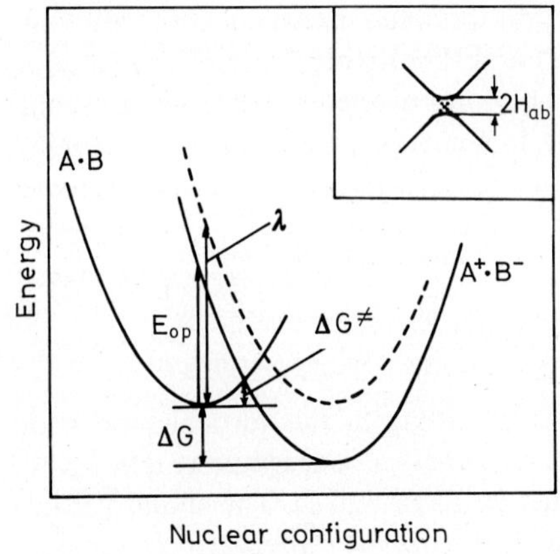

Fig. 1. Profile of the potential energy curves of reactants and products of an electron transfer reaction as a function of nuclear configuration

energy of reactants and products as a function of a reaction coordinate made up of an appropriate combination of the nuclear coordinates of the system. These coordinates are of two types: (i) inner, i.e. internal coordinates (bond lengths and angles) of the reacting molecules; (ii) outer, i.e. coordinates specifying the arrangement of the solvent surrounding reactants and products. The vertical displacement of the minima is related to the energetics of the reaction, while the horizontal displacement represents the fact that reactants and products certainly have different equilibrium solvation shells, and often also correspond to different equilibrium molecular geometries. In order for the Franck-Condon principle to be obeyed, a distortion in the outer and inner nuclear coordinates of the reactants leading to a geometry where reactants and products are isoenergetic (crossing point of the two surfaces) is required prior to electron transfer. Of all the crossing points and distortional pathways available in the multidimensional nuclear space of the system, those of lowest energy correspond to the transition state and the reaction coordinate, respectively, of the electron transfer reaction (Fig. 1).

Thus, the activation free energy $\Delta G^{\neq}$ in Eq. 2 corresponds (converting from energies to free energies) to the energy difference between the crossing point and the reactant minimum in Fig. 1. In view of its physical meaning, the activation term in Eq. 2 can also be called the *nuclear* term of the rate expression. The k_{el} term in Eq. 2 is the transmission coefficient of the reaction, i.e. the probability that the reactants, when reaching the geometry of the crossing point, convert into products. Owing to its physical origin (vide infra), k_{el} can also be called the *electronic* factor of the rate constant. In Eq. 2, ν_N is the nuclear frequency factor of the reaction, which sets the maximum possible value for the rate constant. It can be expressed [45] as a weighted mean of the frequencies of the various nuclear modes involved in the reaction coordinate. As such, it tends to be dominated by the high-frequency inner modes (typical values, 4.5×10^{13} s^{-1} for C—C stretching of aromatic ligands and $(0.9\text{–}1.5) \times 10^{13}$ s^{-1} for metal-ligand stretching in coordination compounds).

It is evident from Fig. 1 that the activation free energy is determined by the combined effect of the degree of distortion between products and reactants (horizontal displacement of the two curves) and the driving force of the reaction (vertical displacement of the two curves). The Marcus theory expresses this combined dependence in terms of a parabolic free-energy relationship [42] (Eq. 3). In Eq. 3, ΔG is the standard free energy change of the reaction. The λ parameter

$$\Delta G^{\neq} = (\lambda/4)\left(1 + \frac{\Delta G}{\lambda}\right)^2 \tag{3}$$

is the so-called *reorganizational energy*, corresponding to the vertical separation between reactant and product curves for a hypothetical isoergonic reaction with the same nuclear distortions (dashed product curve in Fig. 1). The actual vertical energy difference corresponds to the energy of the optical "intervalence transfer" transition to be discussed in Sect. 2.2. The reorganizational energy can be split into the sum of two independent contributions

$$\lambda = \lambda_{in} + \lambda_{out} \tag{4}$$

which can be calculated from expressions of the Marcus theory [42–46], provided that the appropriate parameters (structural data and vibrational frequencies of the reacting subunits and solvent dielectric constants) are known. Alternatively, the reorganizational energy for the reaction can be obtained as the arithmetic mean of the experimental activation energies of the corresponding "self-exchange" reactions of the A/A^+ and B/B^- couples [42]. Qualitatively, the outer part of the reorganizational energy increases with increasing solvent polarity and A—B distance. The inner part depends on the degree of geometrical distortion (for coordination compounds, mainly metal-ligand bond length changes) occurring in the

A/A^+ and B/B^- couples. For octahedral metal complexes, small λ_{in} values are expected for couples that involve t_{2g} redox orbitals (such as, e.g., Ru(III)/Ru(II), Os(III)/Os(II) or Re(II)/Re(I)), and large λ_{in} values for couples involving e_g^* redox orbitals (such as, e.g. Co(III)/Co(II), Rh(III)/Rh(II), or Cr(III)/Cr(II)). Reductions occurring at largely delocalized organic ligands (e.g. polypyridine systems) are expected to have small associated reorganizational energies.

A point that has attracted a considerable deal of experimental and theoretical attention is the behavior predicted by the Marcus free energy relationship (Eq. 3) in the highly exergonic ΔG region. According to Eq. 3, $\Delta G^{\neq}$ equals $\lambda/4$ at $\Delta G = 0$, goes to 0 at $\Delta G = -\lambda$, and increases again for more negative ΔG values. This can be easily visualized by thinking at how the crossing point shifts when the exergonicity of the reaction is increased (Fig. 2): from the right branch (Fig. 2a, "normal" activated process) through the minimum (Fig. 2b, activationless process) to the left branch (Fig. 2c, "inverted" activated process) of the reactant curve. Therefore, for moderately exergonic reactions the driving force is expected to help the reaction kinetics, but for strongly exergonic reactions the driving force is predicted to act against it. The ΔG region in which this intuitively odd effect is expected ($\Delta G < -\lambda$) is usually indicated as the "Marcus inverted region". This feature is also predicted by quantum mechanical models of electron transfer processes [47]. After a long period of experimental search and many unsuccessful attempts [48, 49], the prediction of the inverted region is now supported by definite experimental evidence [50, 51].

The transmission coefficient k_{el} in Eq. 2 is related to the detailed shape of the potential energy curves in the intersection region. Strictly speaking, the reactant and product potential energy curves of Fig. 1 correspond to zero-order wavefunctions of the system. If there were no electronic interaction between these zero-order states, no mechanism for transition from reactants to products would be available. Actually, in most practical systems a small but finite electronic inter-

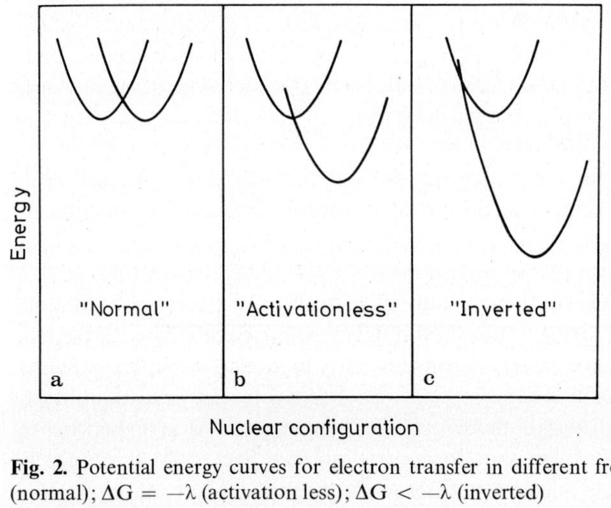

Fig. 2. Potential energy curves for electron transfer in different free energy regions: $\Delta G > -\lambda$ (normal); $\Delta G = -\lambda$ (activation less); $\Delta G < -\lambda$ (inverted)

action occurs between A and B in the reactant pair, and a perturbation hamiltonian H_{ab} coupling the initial (A.B) and final ($A^+.B^-$) states of the system should be considered. This electronic coupling mixes the zero-order states in the intersection region, leading to the first-order avoided crossing surfaces shown in the inset of Fig. 1. A quantitative expression for the transmission coefficient can be obtained [45] within the framework of the Landau-Zener treatment of avoided crossings. The relevant result is

$$k_{el} = \frac{2[1 - \exp(-\nu_{el}/2\nu_N)]}{2 - \exp(-\nu_{el}/2\nu_N)} \tag{5}$$

$$\nu_{el} = \frac{2H_{ab}^2}{h}\left(\frac{\pi^3}{\lambda RT}\right)^{1/2} \tag{6}$$

Two limiting cases can be identified [45] on the basis of Eqs. 2, 5, 6:

(i) If the electronic interaction H_{ab} is very small, $\nu_{el} \ll 2\nu_N$, $k_{el} = (\nu_{el}/\nu_N) \ll 1$, and

$$k = \nu_{el} \exp(-\Delta G^{\neq}/RT) \tag{7}$$

This is called the "nonadiabatic" limit of electron transfer reactions, in which the rate determining step is electron transfer at the transition state geometry. The unimolecular rate constant is much smaller than the nuclear frequency and is very sensitive to factors that may influence the degree of electronic interaction between the reactants (e.g. center-to-center distance, steric hindrance of substituents, orientational factors, nature of interposed groups or medium, etc.).

(ii) If H_{ab} is sufficiently high that $\nu_{el} \gg 2\nu_N$, $k_{el} = 1$ and

$$k = \nu_N \exp(-\Delta G^{\neq}/RT) \tag{8}$$

This is called the "adiabatic" limit of electron transfer reactions, in which the rate determining step is the nuclear motion that leads to the transition state geometry. The unimolecular reaction rate constant may approach (for small $\Delta G^{\neq}$) the nuclear frequency factor, and the reaction is insensitive to factors that may influence the degree of electronic interaction between the reactants.

The value of H_{ab} depends on the overlap between the electronic wavefunctions of the donor and acceptor groups, that should decrease exponentially with increasing donor-acceptor distance. Calculations of H_{ab} in actual complex systems, particularly for coordination compounds, are prohibited by the unavailability of reliable wavefunctions, although there may be ways to arrive at experimental estimates of its magnitude from spectroscopic data (see Sect. 2.2). It should be noticed that the amount of electronic interaction required to promote electron transfer is very small in a common chemical sense. In fact, it can be easily verified by putting reasonable numbers in Eqs. 2, 5, 6 that, for an activationless reaction, H_{ab} values of a few wavenumbers are sufficient to give rates in the subnanosecond time scale, and a few hundred wavenumbers are sufficient to reach the limiting adiabatic regime.

The Marcus model, as outlined above, refers to a pair of A and B molecular reactants (e.g. two mononuclear transition metal complexes) at fixed ("contact") distance. The same basic model can be applied to describe intramolecular electron transfer between two metal centers in a binuclear complex, as schematized in Eq. 9, where M_a and M_b represent two

$$M_a—B—M_b \rightarrow M_a^+—B—M_b^- \tag{9}$$

metal-containing fragments and B is a bridging ligand. The obvious difference is that the distance between the centers and the relative geometry of the donor and acceptor fragments is now fixed by the presence of the bridging ligand (except for the case of flexible bridges). The extension of the model to this situation, however, requires some comment. As compared with an analogous intermolecular reaction at the same center-to-center distance, the reorganizational energy (and thus the nuclear part of the rate constant) is not expected to be drastically altered by the presence of the bridging ligand. On the contrary, relatively important effects of the bridging ligand are expected to occur on the electronic part of the rate constant. In fact, depending on its length and electronic structure, the bridging ligand can induce a more or less important degree of delocalization between the metal centers, thus increasing H_{ab} with respect to the corresponding intermolecular value at the same center-to-center distance. The role of the bridging ligand in enhancing the electronic coupling between metal centers in a polynuclear complex has been described by a number of authors [52–54] in terms of "superexchange", i.e. configuration interaction between the initial ($M_a—B—M_b$) and final ($M_a^+—B—M_b^-$) zero-order states of the electron transfer process and high-energy charge transfer states involving the bridging ligand, such as $M_a^+—B^-—M_b$ and $M_a—B^+—M_b^-$. The relevance of the magnitude of H_{ab} to the problem of localization vs delocalization in polynuclear complexes will be dealt with in some detail in Sect. 2.4.

2.2 Optical and Photoinduced Electron Transfer

The Marcus model makes it clear that reactants and products of an electron transfer process are intertwined by a ground/excited state relationship. For example, for nuclear coordinates that correspond to the equilibrium geometry of the reactants, $A^+.B^-$ is an electronically excited state of A.B (Fig. 1). In principle, as a consequence, optical transitions connecting the two states are possible, as indicated by the vertical arrow in Fig. 1. These *optical* electron transfer transitions are usually denominated "intervalence transfer" (IT) transitions. The Hush theory [43, 45, 55] correlates the spectroscopic features of the IT transition to the parameters that are involved in the corresponding thermal electron transfer process by means of Eqs. 10–12,

$$E_{op} = \lambda + \Delta G \tag{10}$$

$$\Delta\bar{\nu}_{1/2} = 48.06(E_{op} - \Delta G)^{1/2}\ \mathrm{cm}^{-1} \tag{11}$$

$$\varepsilon_{max}\Delta\bar{\nu}_{1/2} = H_{ab}^2 \frac{r(\text{Å})^2}{4.20 \times 10^{-4}\ E_{op}} \tag{12}$$

where E_{op}, $\Delta\bar{\nu}_{1/2}$, and ε_{max} are the energy, halfwidth, and maximum intensity of the IT band, and r is the center-to-center distance. As shown by Eqs. 10–12, the energy depends on both reorganizational energy and thermodynamics, the halfwidth reflects the reorganizational energy, and the intensity of the IT transition is mainly related to the magnitude of the electronic coupling between the two redox centers.

In principle, therefore, important kinetic information on a thermal electron transfer process could be obtained from the study of the corresponding optical IT transition. For bimolecular reactions, the observation of the IT transitions is prevented by the exceedingly low concentration of A.B encounter pairs in solution, unless A and B are oppositely charged species and form ion pairs (in such a case the observed IT transitions are labelled as "outer-sphere"). The situation is obviously much more favourable in the case of intramolecular electron transfer within covalently linked donor-acceptor pairs, such as e.g. polynuclear complexes. In most practical cases, for isoergonic reactions the optical transitions usually lie in the near infrared, whereas for substantially endoergonic reactions the IT bands can appear in the visible or ultraviolet regions (for exergonic reactions, the statement should be referred to the "reverse" IT transition $A^+.B^- \rightarrow A.B$). In practice, due to the dependence of the intensity on H_{ab}, IT transitions may only be observed in systems with relatively strong inter-component electronic coupling (e.g. for H_{ab} values of 10, 100, and 1000 cm^{-1}, ε_{max} values of 0.2, 20, and 2000, respectively, are obtained from Eq. 12 using $E_{op} = 15000\ \mathrm{cm}^{-1}$, $\Delta\bar{\nu}_{1/2} = 4000\ \mathrm{cm}^{-1}$, and $r = 7$ Å). By recalling the figures given in Sect. 2.1, it is clear that weakly coupled systems may undergo relatively fast electron transfer processes without exhibiting appreciably intense IT transitions.

In principle, besides IT absorption, the possibility of IT emission should also be considered. Inspection of Fig. 2 clearly shows that emission from a relaxed state is a meaningful possibility only in the case of an electron transfer process in the inverted region. Emissions of this type have recently been reported for some covalently bound organic donor-acceptor systems [56]. On the other hand, no observation of such processes in polynuclear transition metal complexes seems to have been reported until now.

Optical (IT) electron transfer should not be confused with *photoinduced* electron transfer. This term is generally used to describe a thermal electron transfer process that follows electronic excitation of one of the reaction partners (e.g. Eqs. 13, 14).

$$A.B \xrightarrow{h\nu} A^*.B \tag{13}$$

$$A^*.B \rightarrow A^+.B^- \tag{14}$$

The relationship between thermal, optical, and photoinduced electron transfer is illustrated in Fig. 3. In Sect. 3, examples of both optical and photoinduced electron transfer in polynuclear complexes will be discussed.

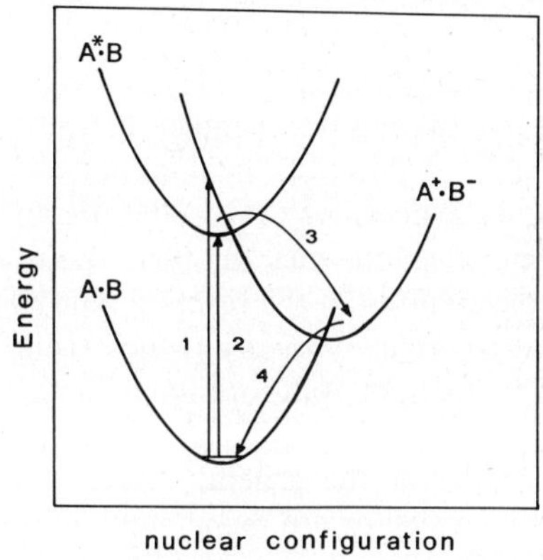

Fig. 3. Schematic representation of different type of electron transfer processes in a two-center system: (*1*) optical, (*2* + *3*) photoinduced, (*4*) thermal

2.3 Energy Transfer

Electronic energy transfer [57] is a process by which simultaneous deactivation of an excited molecule and excitation of a ground-state molecule occurs (Eq. 15)

$$A^*.B \rightarrow A.B^* \tag{15}$$

The energy transfer process must obey energy conservation (which means that B* must be equal or lower in energy than A*), and requires some kind of electronic interaction between the donor and the acceptor. Following standard arguments [58], the electronic interaction between two molecular species can be split into two additive terms, a *coulombic* term and an *exchange* term. The two terms have different dependences on various parameters of the system (spin of ground and excited states, donor-acceptor distance, etc.) and each of them can become predominant depending on the specific system and experimental situation. This leads to the identification of two main energy transfer mechanisms.

The coulombic (also called "resonance" or "Forster-type") mechanism [57] is a long-range mechanism that does not require physical contact between donor and acceptor. It can be shown that the most important term within the coulombic interaction is the dipole-dipole term, that obeys the same selection rules as the corresponding electric dipole transitions of the two partners ($A^* \rightarrow A$ and $B \rightarrow B^*$). Therefore, coulombic energy transfer is expected to be efficient in systems in which the radiative transitions connecting the ground and the excited state of each partner have high oscillator strength. Thus, the typical example of efficient coulombic mechanism is that of singlet-singlet energy transfer ($A^*(S_1).B(S_0) \rightarrow A(S_0).B^*(S_1)$) between large aromatic molecules, a process used by nature in the "antenna" part of the photosynthetic apparatus [59]. As a general rule, in coordination compounds the only excited state of appreciable lifetime is the lowest excited state [60], and the transition from this state to the ground state is spin-forbidden. Therefore, with coordination compounds as donors and acceptors, coulombic energy transfer is not expected to be a widespread phenomenon (see, however, Sect. 3.1 for some exceptions to this statement).

The exchange (also called Dexter-type) mechanism [57] is a short-range mechanism that requires orbital overlap, and therefore physical contact, between donor and acceptor. The exchange interaction can be visualized as the simultaneous exchange of two electrons between the donor and the acceptor via LUMOs (from A to B) and HOMOs (from B to A). The spin selection rules for this type of mechanism arise from the need to obey spin conservation in the reacting pair as a whole. This allows this mechanism to be operative in many cases in which, as in coordination compounds, the relevant excited states are spin-forbidden in the usual spectroscopic sense.

The above mechanistic considerations are made for the general case of an A*.B pair of molecules. When the donor and the acceptor are metal-containing components of a binuclear complex, e.g. $M_a^*—B—M_b$, the presence of the bridging ligand is not expected to change substantially the situation as far as a coulombic mechanism (at the same center-to-center distance) is concerned. The situation may be different, on the other hand, for an exchange mechanism. It has been pointed out in Sect. 2.1 and 2.2 that the presence of a bridging ligand can increase the electronic coupling matrix element for electron transfer in a binuclear complex relative to that of an analogous bimolecular reaction. Although the matrix elements involved in the two types of processes are somewhat different [61], the concept that electron delocalization via the bridging ligand can increase the interaction can be extended from electron to exchange energy transfer. Examples of very efficient exchange energy transfer in polynuclear complexes will be discussed in particular in Sect. 3.5.

An interesting question is that concerning the relative rates of electron and exchange energy transfer in an ideal system in which both processes were thermodynamically allowed. No general answer to this question is available, but a few points can be stressed. To discuss this question, a kinetic model of energy transfer [18] can be used that considers electronic and nuclear factors in the rate constant of energy transfer, much in the same way as the Marcus model does for electron transfer. Generally speaking, energy transfer tends to have smaller reorganizational barriers than electron transfer, due to the much smaller solvent repolarization required for the former process. As far as the electronic term is concerned, the two-electron vs one-electron nature of the interaction intuitively implies more severe overlap requirements for energy than for electron transfer. In a number of elegant studies on covalently linked organic donor-acceptor systems, Closs and Miller [62] have recently probed the relationship between electron and exchange energy transfer, showing that rates of energy transfer decrease with increasing bridge length much faster than those of corresponding electron transfer processes.

2.4 Polynuclear Complexes as Supramolecular Systems

In Sects. 2.1–2.3, models for intermolecular electron and energy transfer have been briefly described, and it has been pointed out that the same models can be used, with minor adjustments, to describe intramolecular electron and energy transfer processes in covalently linked donor-acceptor systems, such as, e.g. $M_a—B—M_b$ (Eqs. 16–18)

$$M_a—B—M_b \rightarrow M_a^+—B—M_b^- \quad \text{thermal electron transfer} \quad (16)$$

$$M_a^*—B—M_b \rightarrow M_a^+—B—M_b^- \quad \text{photoinduced electron transfer} \quad (17)$$

$$M_a^*—B—M_b \rightarrow M_a—B—M_b^* \quad \text{energy transfer} \quad (18)$$

It must be realized, however, that this extension relies on the assumption that, for a $M_a—B—M_b$ binuclear complex, the very concepts of energy and electron transfer retain their meaning. This is not a trivial point, as it involves assumptions on the degree of electronic delocalization in the polynuclear complex. It is indeed intuitively clear that the concepts of intramolecular energy and electron transfer would become meaningless if the $M_a—B—M_b$ species were a completely delocalized molecule. In other words, in order to consider energy and electron transfer within a polynuclear complex in the same sense as one does for an intermolecular process, the degree of electronic delocalization between the M_a and M_b centers must be sufficiently small that:

(i) integral oxidation states can be assigned to the M_a and M_b metal centers;
(ii) independent excitation of the M_a and M_b subunits is feasible.

The first point has been discussed in considerable detail as a well-defined borderline in mixed valence chemistry [55, 63, 64]. Let us consider the potential energy curves for an isoenergetic electron transfer process within a binuclear complex $M_a—B—M_b$ (e.g. $M_b = M_a^+$) (Fig. 4), using the same physical model and the same parameters as described in Sects. 2.1 and 2.2. The problem of localization can be easily discussed in terms of the relative magnitudes of H_{ab} and λ. The curves in Fig. 4a correspond to the hypothetical case of no interaction between the metal centers ($H_{ab} = 0$). In this case, integral oxidation states for the two metals are clearly appropriate, and one full electronic charge would be transferred in the hypothetical electron transfer process. This is usually called Class I mixed-valence behavior [63]. If, as is usually the case, the metal-metal interaction is not negligible, the crossing between the two zero-order curves is avoided and a $2H_{ab}$ separation between the lower and upper curve arises (Fig. 4b). In Figure 4b, H_{ab} is clearly

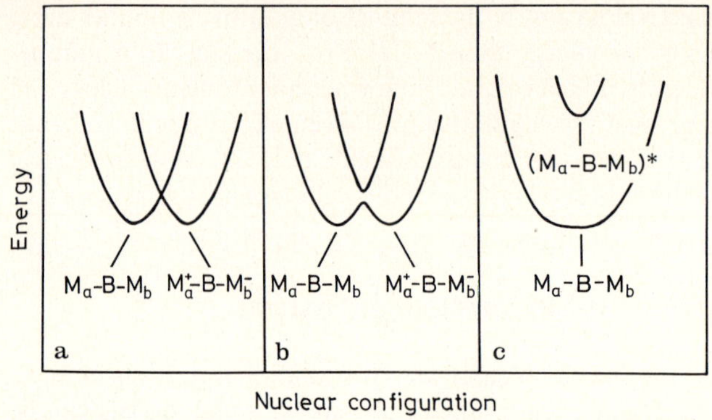

Fig. 4. Potential energy curves for a symmetric mixed-valence complex with different relative magnitudes of λ and H_{ab}

much smaller than λ and the zero-order description involving integral charges on the reaction partners is practically exact everywhere except for a very narrow region of the nuclear space. This is a case in which the system can be fairly considered as localized, and integral oxidation states can be assigned to M_a and M_b in reactants and products. This is usually called Class II mixed-valence behavior [63]. In the case of H_{ab} equal or larger than λ, at the other extreme, a single minimum for both curves would be obtained at the geometry of the zero-order crossing point (Fig. 4c), and the two curves would describe ground and excited states of a fully delocalized system in which nonintegral oxidation states would be appropriate for the M_a and M_b units. This is usually called Class III mixed-valence behavior [63]. Which of these limits is actually encountered in a practical system can be decided, in principle, if the relative magnitude of λ and H_{ab} is known. As already pointed out, the Marcus-Hush theory gives the possibility of calculating λ on the basis of intermetallic distance, solvent, and intrinsic properties of M_a and M_b. On the other hand, for H_{ab}, that depends not only on distance but also on the nature (delocalizing or coupling ability) of the bridge, a calculation from first principles is practically impossible. In several instances, the relative magnitude of λ and H_{ab} can be often experimentally evaluated by putting together electrochemical (consecutive potential differences) structural (internal equilibrium coordinates of M_a and M_b) and spectral information (energy, intensity and shape of the intervalence transfer band). In a large majority of polynuclear complexes studied so far, including most of the systems discussed in this review, the relative magnitudes of λ and H_{ab} are such that a *localized* description with integral oxidation states is appropriate.

As for the ground state, an essentially localized description is also adequate for the excited states of most polynuclear complexes. In other words, locally excited states of the metal-containing subunits are present (e.g. M_a^*—B—M_b and M_a—B—M_b^*) that do not significantly differ from those of the component units or of suitable models thereof. This, on the other hand, does not mean that a localized polynuclear complex must *only* have locally excited states. For example, the intervalence transfer states discussed in Sect. 2.2 involve simultaneously two metal-containing subunits and thus belong to the polynuclear complex as a whole. Other types of inter-component charge transfer states will be encountered in later sections. Despite their non-local character, however, such charge transfer states are still based on a localized description, in which excitation involves the transfer of an essentially integral electronic charge between localized redox sites of the polynuclear complex.

To the extent to which polynuclear complexes are amenable to a localized description in the above specified sense, they can be considered as *supramolecular* systems. While a general, clear-cut definition of this term is difficult [33], a supramolecular system is intended here as a complex system made up of molecular subunits (components) with definite individual properties. A supramolecular system may have, in addition to the properties of the component subunits, new (supramolecular) properties characteristic of the ensemble. Although in a M_a—B—M_b polynuclear complex some ambiguity in the definition of the subunits is given by the presence of the bridging ligand [65], it is clear from the above discussion that M_a and M_b (or M_aB and M_bB) are molecular subunits with well defined individual properties. Also, supramolecular properties, such as, e.g. intervalence transfer absorption, can be easily identified in polynuclear complexes.

Finally, an important point should be stressed concerning the nature of the "reactants" and "products" in an intramolecular energy or photoinduced electron transfer process. It is easy to see (Eqs. 17, 18) that in such processes reactants and products are simply different electronically excited states of the system: the reactant is a locally excited state, the product of energy transfer is the other locally excited state, and the product of electron transfer is the intervalence transfer state. From this point of view, intramolecular energy and electron transfer processes could simply be considered as radiationless transitions between excited states of a large molecule. Given the localized nature of polynuclear complexes, however, a supramolecular picture in terms of intercomponent transfer processes seems to be more insightful than the simple "molecular" approach.

3 Survey of Experimental Studies

In this section, experimental studies on the photophysical and photochemical behavior of polynuclear transition metal complexes are reviewed. The survey covers the period 1979–1989, where most of the work in this area has been performed, although a few important earlier studies are also considered.

Given the rather inhomogeneous nature of the studies reviewed, it has been difficult to find a single satisfactory criterion to organise the material. Generally speaking, organisation according to types of behavior has been preferred. Where possible, studies on homogeneous series of complexes have been grouped together.

Except for the simplest ones, ligands are indicated with abbreviations. The reader is referred to the Appendix for a list of abbreviations (Sect. 6.1) and corresponding structural formulae (Sect. 6.2).

3.1 Energy Transfer in Weakly Coupled Polychromophoric Systems: Polypyridine Complexes Linked by Saturated Organic Chains

Studies of polynuclear transition metal complexes in which the metal centers are very weakly coupled are described in this section. The bridging ligands of these complexes (bibpy-n, see Sects. 6.1 and 6.2) are made of two bipyridine-type ligands covalently linked through a saturated hydrocarbon chain. Binuclear and tetranuclear complexes of general formula $(L_a)_2M_a(B)M_b(L_b)_2$ and $[(L_a)_2M_a(B)]_3M_b$, respectively, where L_a, L_b are nonbridging polypyridine ligands, M_a, M_b are metals centers and B is the bridging ligand, are discussed. Because of the saturated hydrocarbon link, the two bipyridine moieties of this bridging ligand behave in a totally independent way. Thus, for these polynuclear complexes it is easy to identify good mononuclear analogs of the components, e.g. $M_a(L_a)_2(Me_2\text{-bpy})$ and $M_b(L_b)_2(Me_2\text{-bpy})$, and $M_b(Me_2\text{-bpy})_3$. As far as the photophysical behavior of the mononuclear complexes is concerned, the Ru(II) and Os(II) complexes have metal-to-ligand charge transfer (MLCT) states of formally triplet character as the lowest excited state. These states are long lived and emit efficiently in room temperature fluid solution. The Fe(II) complexes, on the other hand, do not emit in fluid solution because of the presence of low lying d—d states.

Polynuclear complexes with a bridging ligand that contains three covalently linked bipyridine units are also discussed.

The relevant properties of the polynuclear complexes discussed in this section are summarized in Table 1.

3.1.1 Symmetrical Homometallic Complexes

The homometallic binuclear complex $(Me_2\text{-bpy})_2Ru(\text{bibpy-8}')Ru(Me_2\text{-bpy})_2^{4+}$ has been studied by Schmehl and coworkers [66], while analogous complexes of the type $(bpy)_2Ru(B)Ru(bpy)_2^{4+}$ have been studied by Furue et al., [67, 68] (B = bibpy-2, bibpy-3, bibpy-3′) and Rillema and coworkers [69] (B = bibpy-2). The spectroscopic and electrochemical properties of all these binuclear complexes are almost identical to those of the component monomers (Table 1). For $(Me_2\text{-bpy})_2Ru(\text{bibpy-8}')Ru(Me_2\text{-bpy})_2^{4+}$ the comproportionation equilibrium (Eq. 19) has $K_c < 7\ M^{-1}$ and no intervalence transfer band can be observed in the

$$\text{Ru(II)–Ru(II)} + \text{Ru(III)–Ru(III)} \underset{}{\overset{K_c}{\rightleftarrows}} 2\ \text{Ru(II)–Ru(III)} \tag{19}$$

mixed-valence form [66]. The MLCT emission maxima, lifetimes, and luminescence quantum yields of all these symmetrical binuclear complexes are practically coincident with those of the parent monomers. These results clearly indicate that the two metal centers do not interact with each other, as a consequence of the insulating character of the hydrocarbon chains.

A puzzling behavior is exhibited by the symmetrical tetranuclear complex $[(Me_2\text{-bpy})_2Ru(\text{bibpy-8}')]_3Ru^{8+}$ studied by Schmehl and coworkers [66]. For this complex the spectroscopic and redox properties are very similar to those of the parent monomer $Ru(Me_2\text{-bpy})_3^{2+}$. On the other hand, the MLCT luminescence quantum yield is only 60% of that of the corresponding monomer, while the measured lifetime is the same. As a tentative explanation of the observed intensity quenching, the authors proposed that the tetranuclear complex may have a smaller radiative rate constant with respect to the mononuclear one.

Interesting complexes of the type $[(bpy)_2Ru]_n(tribpy)^{2n+}$ (n = 1–3) have been reported by Balzani and coworkers [70], using the "tripod" tris-bipyridine tribpy ligand to coordinate one, two or three $Ru(bpy)_2^{2+}$ units. The complexes have spectroscopic, photophysical and electrochemical properties that are independent of the number of Ru(II) units, indicating that there is no interaction between the various metal centers. The free tripod ligand exhibits an intense luminescence at 77 K. This emission is not present in the trinuclear complex, where all the bpy arms of tribpy are coordinated to $Ru(bpy)_2^{2+}$ units, because of conversion to the lowest emitting MLCT state taking place within each Ru(II)-based chromophoric unit. In the mononuclear $Ru(bpy)_2(tribpy)^{2+}$ complex two bpy-type arms are free, and in principle ligand centered (LC) emission (that comes from excited states that are localized on the single bpy units of the tribpy ligand) should be observable. Actually, no LC luminescence was found in this complex. Furthermore, excitation in the LC absorption region of two equally absorbing solutions of the mononuclear and trinuclear complexes results in a MLCT emission of comparable intensity. Thus, efficient intramolecular energy transfer from the non coordinated arms of tribpy to the metal containing unit occurs in the mononuclear complex [70].

3.1.2 Heterometallic Complexes

The binuclear $(bpy)_2Ru(bibpy\text{-}2)PtCl_2^{2+}$ complex has been reported by Rillema and coworkers [69]. For this complex, the visible absorption spectrum is coincident with the sum of those of the corresponding mononuclear species ($Pt(bpy)Cl_2$ and $Ru(bpy)_2(bibpy\text{-}2)^{2+}$), and two reversible one-electron reduction processes at potentials close to those of the corresponding mononuclear species are observed. These observations are consistent with the insulating effect of the hydrocarbon chain of the bridging ligand. Spectroscopic and redox properties show that there is no appreciable ground-state interaction between the ruthenium- and the platinum-containing moieties. As far as the photophysical behavior is concerned, the emission is very similar in lifetime and quantum yield to that of the corresponding Ru-based component. This indicates that the Pt-based fragment has no quenching effect, presumably because it lacks excited-states at lower energy.

The interesting binuclear complex $(bpy)_2Ru(bibpy\text{-}3')Os(bpy)_2^{4+}$ has been studied by Furue et al. [71]. In this complex, selective excitation of the two fragments is not possible due to spectral overlap, so that the emission spectrum consists of Ru-based emission at higher energy and Os-based emission at lower energy. However, with respect to the emission of the corresponding mononuclear species, a decrease of the intensity of the Ru-based emission and a substantial enhancement of the emission of the osmium moiety are observed. Since no quenching process is observable in 1:1 mixtures of the corresponding monomers at comparable concentration, the authors conclude that efficient intramolecular quenching of the Ru-based emission takes place in the binuclear complex [71]. On

Table 1. Spectroscopic, Photophysical and Redox Properties of the Complexes[a]

Complexes	Absorption[b] λ_{max}, nm (ε^{d})	Emission λ_{max}, nm (τ, μs) 298 K	Emission 77 K	Redox Properties[c] $E^{ox}_{1/2}$, V	$E^{red}_{1/2}(1)$, V	$E^{red}_{1/2}(2)$, V	Ref.
mononuclear							
$Ru(bpy)_3^{2+}$	452 (13)	615 (1.10)	580 (5.00)	+1.28	−1.32	−1.52	24
$Ru(Me_2\text{-}bpy)_3^{2+}$	459 (14.1)	621 (1.06)	595 (3.99)	+1.10	−1.46	−1.63	66
$Ru(dec\text{-}bpy)_3^{2+}$	467 (25)	629 (2.23)	607	+1.55	−0.91	−1.05	72
$Ru(biq)_3^{2+}$	524 (9)	705[e]		+1.47	−0.73		24
$Ru(Me_2\text{-}bpy)_2(dec\text{-}bpy)^{2+}$	493 (13)	695 (0.85)	615	+1.30	−1.03	−1.53	72
$Ru(Me_2\text{-}bpy)_2(bibpy\text{-}8')^{2+}$	461 (14.1)	617 (0.82)	595 (4.38)	+1.11	−1.46	−1.65	66
$Ru(biq)_2(Me_2\text{-}bpy)^{2+}$	550 (6.5)	750	739 (2.20)	+1.35	−0.84	−1.08	73
$Ru(bpy)_2(bibpy\text{-}8')^{2+}$	456 (14.1)	621	586 (4.15)	+1.23	−1.37	−1.56	73
$Ru(bpy)_2(bibpy\text{-}2)^{2+}$		610 (0.356)[f]	(4.56)				72
$Ru(bpy)_2(bibpy\text{-}5)^{2+}$		610 (0.390)[f]	(4.09)				72
$Ru(bpy)_2(bibpy\text{-}12)^{2+}$		610 (0.360)[f]	(4.07)				72
$Ru(bpy)_2(Me_2\text{-}bpy)^{2+}$	455 (13)[g]	613 (0.49)[g]		+1.22	−1.37	−1.57	68
$Ru(Me_2\text{-}bpy)(dec\text{-}bpy)_2^{2+}$	483 (16.2)	658 (0.87)	633 (5.30)	+1.44	−0.96	−1.16	66
$Os(bpy)_2(Me_2\text{-}bpy)^{2+}$	590[h] (4) 650[h] (4)	720 (0.018)[g]		+0.77			71
$Pt(bpy)Cl_2$					−1.31[i]		69
binuclear							
$(biq)_2Ru(bibpy\text{-}8')Ru(bpy)_2^{4+}$	456 (17) 550 (9.4)	620 748	586 (0.30) 739 (2.30)	+1.22 +1.35	−0.85	−1.08	73
$(Me_2\text{-}bpy)_2Ru(bibpy\text{-}8')Ru(Me_2\text{-}bpy)_2^{4+}$	461 (27.5)	617 (0.82)	602 (4.19)	+1.11	−1.46	−1.65	66
$(Me_2\text{-}bpy)_2Ru(bibpy\text{-}8')Ru(dec\text{-}bpy)_2^{4+}$	463 (21.4)	658 (1.14)	633 (5.31)	+1.11 +1.42	−0.96	−1.15	66
$(bpy)_2Ru(bibpy\text{-}2)Ru(bpy)_2^{4+}$	455[g]	614[g] (0.630[g])		+1.21	−1.38		67
$(bpy)_2Ru(bibpy\text{-}3)Ru(bpy)_2^{4+}$	455[g]	616[g] (0.520)[g]		+1.21	−1.38		67
$(bpy)_2Ru(bibpy\text{-}3')Ru(bpy)_2^{4+}$	455[g]	615[g] (0.481)[g]		+1.21	−1.38		67

$(bpy)_2Ru(bibpy\text{-}3')Os(bpy)_2^{4+}$	455[h]	615[f] (0.002)[g]		+0.77			71
	590[h] (4)	720[f] (0.018)[g]		+1.21			
	650[h] (4)						
$(bpy)_2Ru(bibpy\text{-}2)PtCl_2^{2+}$	457[i] (13)	620 (1.03)[i]		+1.19[i]	−1.27[i]	−1.38[i]	69
tetranuclear							
$[(Me_2\text{-}bpy)_2Ru(bibpy\text{-}8')]_3Ru^{8+}$	461 (60.3)	621 (0.77)	599 (4.10)	+1.11	−1.46	−1.63	66
$[(dec\text{-}bpy)_2Ru(bibpy\text{-}8')]_3Ru^{8+}$	483 (72.4)	662 (0.97)	641 (4.74)	+1.11	j	j	66
				+1.42			
$[(bpy)_2Ru(bibpy\text{-}8')]_3Fe^{8+}$	525[f] (7.4)	615[f] (0.038)[f]	(0.10)				72
$[(bpy)_2Ru(bibpy\text{-}2)]_3Fe^{8+}$	525[f]	610[f] (0.006)[f]	(0.045)				72
$[(bpy)_2Ru(bibpy\text{-}5)]_3Fe^{8+}$	525[f]	610[f] (0.052)[f]	(0.11)				72
$[(bpy)_2Ru(bibpy\text{-}12)]_3Fe^{8+}$	525[f]	610[f] (0.092)[f]	(0.085)				72

[a] All data are for deaerated CH_3CN solutions unless otherwise noted; [b] low-energy MLCT bands; [c] potentials vs SCE; [d] 10^{-3} M^{-1} cm^{-1}; [e] immeasurably short; [f] water/methanol 1:1, [g] water; [h] methanol; [i] propylene carbonate; [j] no distinct waves appear in CV due to extensive overlap of neighboring reductions.

thermodynamic grounds, both electron transfer (Eq. 20) or energy transfer (Eq. 21) quenching mechanisms are possible (in Eqs. 20 and 21 the M → bpy MLCT

$$(bpy)(bpy^{-\cdot})Ru^{III}(B)Os^{II}(bpy)_2^{4+} \rightarrow (bpy)(bpy^{-\cdot})Ru^{II}(B)Os^{III}(bpy)_2^{4+} \quad (20)$$

$$(bpy)(bpy^{-\cdot})Ru^{III}(B)Os^{II}(bpy)_2^{4+} \rightarrow (bpy)_2Ru^{II}(B)Os^{III}(bpy^{-\cdot})(bpy)^{4+} \quad (21)$$

excited states are indicated with the $M^{III}(bpy^{-\cdot})$ notation). A careful examination of the time dependence of the emission spectra shows that the decay of the Ru-based emission and the rise of the Os-based emission coincide [71]. This definitely shows that the quenching mechanism is energy transfer. The ratio of the preexponential factors of the two time-dependent processes indicates that the intramolecular energy transfer process has a unitary efficiency. As to the mechanism of this intramolecular energy transfer, the authors point out the following experimental facts: i) there is no experimentally measurable dependence of the energy transfer rate on solvent viscosity; ii) the absorption of the osmium moiety overlaps substantially the emission of the ruthernium moiety. These results suggest a dipole-dipole (Förster type) mechanism for the intramolecular energy transfer process. This suggestion is confirmed by the agreement between the experimental energy transfer rate constant ($4.7 \times 10^8\ s^{-1}$) and the value calculated according to Förster model [71].

An interesting series of heterotetranuclear complexes of general formula $[(bpy)_2Ru(B)]_3Fe^{8+}$ with B = bibpy-2, bibpy-5, bibpy-12, bibpy-8′, has been studied by Schmehl et al. [72]. These tetranuclear complexes, constituted by a central Fe^{2+} ion coordinated to three $(bpy)_2Ru(B)^{2+}$ moieties, were obtained in solution by in situ complexation of $(bpy)_2Ru(B)^{2+}$ with Fe(II). The solutions exhibit emissions with double exponential decays. The fast component of the decay corresponds to the emission from Ru-based units complexed to iron, and the long component comes from excess monomeric Ru(II) complexes present in solution. The reason for the faster decay of the emission of the polynuclear complexes with respect to mononuclear ones is, according to the authors, intramolecular energy transfer from the ruthenium-containing moieties to the iron center [72]. This conclusion is based on the following observations: i) bimolecular quenching does not occur in solutions containing $Ru(bpy)_3^{2+}$ and $Fe(bpy)_3^{2+}$ at comparable concentrations; ii) the quenching behavior mirrors the formation of the tetranuclear species; iii) the relative importance of the fast decay increases with increasing Fe:Ru molar ratio; iv) the luminescence decay of the corresponding homometallic tetranuclear complex, $[(bpy)_2Ru(bibpy\text{-}8')]_3Ru^{8+}$, does not exhibit a short lived-component. The experimental results do not allow to establish definitely the mechanism (dipole-dipole or exchange) of the energy transfer process. No systematic dependence of the energy transfer rate on the length of hydrocarbon chain of the bridge was obtained either in fluid solutions or in frozen matrix.

3.1.3 Unsymmetrical Homometallic Complexes

The binuclear complexes $(biq)_2Ru(bibpy\text{-}8')Ru(bpy)_2^{4+}$ and $(Me_2\text{-}bpy)_2Ru(bibpy\text{-}8')Ru(dec\text{-}bpy)_2^{4+}$ and the tetranuclear complex $[(dec\text{-}bpy)_2Ru(bibpy\text{-}8')]_3Ru^{8+}$

have been studied by Schmehl and coworkers [66, 73]. These homometallic complexes are unsymmetrycal because of the different nonbridging ligands at the various ruthenium centers. The spectroscopic and redox properties are consistent with those expected for polynuclear complexes with noninteracting metal centers.

For the $(Me_2\text{-bpy})_2Ru(bibpy\text{-}8')Ru(dec\text{-}bpy)_2^{4+}$ complex [66], the 77 K and room temperature emissions are very similar to those of $Ru(dec\text{-}bpy)_2(Me_2\text{-}bpy)^{2+}$, while no emission corresponding to the higher-energy $Ru(Me_2\text{-}bpy)_2(bibpy\text{-}8')$-fragment is observable. On the other hand, in a 1:1 mixture of the two mononuclear species at comparable concentrations two distinct emissions are observed. This indicates that efficient quenching of the Ru → Me_2-bpy MLCT state takes place in the binuclear complex. On thermodynamic grounds, the quenching by electron transfer is slightly endergonic ($\Delta G \approx 0.07$ eV), while quenching by energy transfer is exergonic by approximately 0.12 eV. The excitation spectrum at 77 K clearly indicates that the quenching process is accompanied by sensitization of emission from the $Ru(dec\text{-}bpy)_2(bibpy\text{-}8')$-fragment, demonstrating the occurrence of an intramolecular energy transfer process [66].

For the $[(dec\text{-}bpy)_2Ru(bibpy\text{-}8')]_3Ru^{8+}$ tetranuclear complex [66], the photophysical behavior closely resembles that of the above-discussed binuclear complex.

For the binuclear $(biq)_2Ru(bibpy\text{-}8')Ru(bpy)_2^{4+}$ complex, spectroscopic and redox properties clearly indicate that the fragment containing the $Ru(biq)_2^{2+}$ unit is that with the lowest MLCT energy [73]. Overlap with the other MLCT bands (Ru → bpy and Ru → bibpy-8′) prevents selective excitation of this fragment. The emission spectrum of the complex exhibits both Ru → biq and Ru → bpy (or Ru → bibpy-8′) MLCT emissions. However, the Ru → bpy or (Ru → bibpy-8′) emission is strongly quenched with respect to that of the mononuclear $Ru(bpy)_2$-$(bipy\text{-}8')^{2+}$ complex (at 77 K, $\tau = 300$ ns and 4150 ns for the bi- and mononuclear species, respectively). The excitation spectrum of the Ru → biq MLCT emission matches very closely the absorption spectrum. An analysis of the time-dependence of the emission at 13 K shows that the risetime of the Ru → biq emission coincides with the decay of the Ru → bpy (or Ru → bibpy-8′) emission. These results demonstrate that quenching occurs in this systems by intramolecular energy transfer [73]. An important question concerns the mechanism of the energy transfer process: the small spectral overlap between donor emission and acceptor absorption and the spin-forbidden character of the transition involved might suggest an exchange (Dexter type) mechanism but, in view of the partial singlet character of the formally triplet states, a dipole-dipole energy transfer mechanism cannot be ruled out. Actually, the energy transfer rate constant calculated with the dipole-dipole model is in good agreement with the experimental value ($k = 8.9 \times 10^5\ s^{-1}$ at 13 K) [73].

3.2 Perturbation of the Photophysical Properties of Mononuclear Subunits in Polynuclear Systems: Polyimine-Bridged Complexes

The studies discussed in this section give particular emphasis to the problem of the perturbation of the photophysical properties of mononuclear fragments caused by the formation of a polynuclear system.

The polynuclear complexes dealt with here are constituted by metal-containing fragments linked by multidentate aromatic polyimine ligands of various complexity. Schematically, for a binuclear complex the structure is of type $(L_a)_2M_a(B)M_b(L_b)_2$, where L_a and L_b are nonbridging ligand, and B is a polyimine bridging ligand. With one exception (Sect. 3.2.1), in all of the complexes discussed here the first reduction takes place at the bridging ligand, and the lowest excited state is a metal-to-bridge-charge transfer (MBCT) excited state. A common feature of this class of compounds is that in going from mononuclear to binuclear species an anodic shift in the reduction potential is observed due to the energy stabilization of the bridge π^* orbitals upon binucleation. A parallel red shift of the MBCT transition is observed in the absorption as well as in the emission spectrum. The important mechanistic questions are: i) how these effects depend on the nature of the bridging ligand; ii) how this affects the photophysical behavior of the complex. In the studies discussed below, there seems to be a general agreement concerning the answer to question (i). On the other hand, a thorough understanding of the facts that determine the photophysical behavior of these complexes (question (ii)) has not yet been reached (see Sect. 3.2.5).

There are several studies [74–77] dealing with the mutual perturbation of mononuclear fragments through polyimine bridges, that do not involve any photophysics or photochemistry and thus are not discussed in this survey.

3.2.1 Homometallic Binuclear Polyimine Complexes

The relevant properties of the complexes discussed in this section are reported in Table 2.

The binuclear complex $(bpy)_2Ru(bpm)Ru(bpy)_2^{4+}$ has been studied several years ago by Hunziker and Ludi [78], and Dose and Wilson [79]. They reported that, although the monomer $Ru(bpy)_2(bpm)^{2+}$ is luminescent in room-temperature fluid solution, the corresponding binuclear species is not. On the other hand, Gafney and coworkers [80] reported that both the analogous mono and binuclear complexes with the dpp bridging ligand, $(bpy)_2Ru(dpp)^{2+}$ and $(bpy)_2Ru(dpp)$-$Ru(bpy)_2^{4+}$, are luminescent in room temperature fluid solution. The MBCT emission in the dpp-bridged binuclear complex is found to be slightly red shifted, with shorter lifetime and lower intensity with respect that of the mononuclear species. Gafney [80] interpreted the difference in the photophysical behavior of the bpm-bridged and dpp-bridged complexes on the basis of different degrees of perturbation caused by binucleation. Spectral and electrochemical data showed a strong stabilization of the π^*-acceptor orbital of the bpm bridge upon formation of the binuclear complex, consistent with significant metal-metal interaction. On the other hand, a much smaller effect was observed for the dpp complex, indicating a relatively weak metal-metal interaction. This difference was initially ascribed to the planar structure of the coordinated bpm versus the non planar structure of the coordinated dpp. The comparison between the photophysical behavior of these binuclear complexes lead Gafney to propose [80] that, where the two metal centers are weakly coupled (dpp-bridged complex), the binuclear complex is luminescent; on the contrary, in the case of strong interaction (bpm-bridged complex) the emission of the mononuclear complex is lost in the binuclear species. As will be seen later on, this empirical rule has been widely used by other authors to interpret the photophysical behavior of polyimine bridged polynuclear complexes.

A subsequent study by Gafney and coworkers [81] on the binuclear complex $(bpy)_2Ru(ppz)Ru(bpy)_2^{4+}$, containing the planar ppz bridging ligand, showed that in this complex the metal-metal interaction is intermediate between those of the bpm and dpp complexes. On the basis of this result, the authors concluded that not only the degree of planarity but also the electronic structure of the bridge play a fundamental role in determining the degree of metal-metal interaction and hence the emissive properties of such binuclear complexes. The puzzling result is that, in spite of the larger metal-metal interaction, $(bpy)_2Ru(ppz)Ru(bpy)_2^{4+}$ is a more efficient emitter than the analogous dpp complex. The explanation of this behavior is not clear. The authors suggested the hypothesis of a greater rate of radiationless deactivation for the dpp complex (which, however, is not supported by the reported lifetimes of the two complexes) [81].

The binuclear $(bpy)_2Ru(bpm)Ru(bpy)_2^{4+}$ and $(bpy)_2Ru(dpp)Ru(bpy)_2^{4+}$ complexes have been reexamined together with analogous dpq-bridged complexes by Petersen in recent articles [82, 83], where particular emphasis is placed on the problem of the metal-metal communication across the bridge. Petersen confirmes Gafney's results and interpretation, and examines in detail electrochemical data. Petersen uses the shift in the oxidation potential of the binuclear species with respect to the mononuclear one and the constant K_c of the comproportionation equilibrium (Eq. 22) as a measure of the communication between the metal centers.

$$\mathrm{Ru(II)-Ru(II) + Ru(III)-Ru(III) \overset{K_c}{\rightleftarrows} 2\,Ru(II)-Ru(III)} \tag{22}$$

The comparison of these data for the various binuclear complexes gives ambiguous results. In particular in the case of bpm-bridged binuclear complex the first oxidation potential is substantially more positive with respect to that of the mononuclear one, reflecting a good communication, whereas in the dpp and dpq cases the differences are much smaller, consistent with a weak metal-metal interaction. On the other hand, the comproportionation constants are approximately the same for all of these complexes, suggesting a comparable extent of metal-metal communication, in sharp contrast not only with the previous result but also with the photophysical data.

The complex, $Ru(tpy)(tpp)Ru(tpy)^{4+}$ has been studied by Petersen [83]. For this complex, the substantial shift in the oxidation potential with respect to the analogous mononuclear complex and the large K_c value (larger than that of the bpm, dpp, and dpq binuclear complexes), strongly suggest a good metal-metal interaction. On the other hand, this complex was found to emit in room temperature fluid solution. This behavior is a remarkable exception to the empirical rule of Gafney [80, 81] that a strong metal-metal interaction causes loss of emission. Petersen tentatively ascribed this anomalous behavior to the structure of the tpp tridentate bridging ligand. He observed a puzzling behavior also for the analogous

Table 2. Spectroscopic, Photophysical and Redox Properties of Mono- and Polynuclear Ru(II) Polyimine Complexes[a]

Complexes	Absorption[b] λ_{max}, nm	Emission λ_{max}, nm (τ, ns)		Redox Properties[c] $E^{ox}_{1/2}$, V	$E^{red}_{1/2}$, V	Ref.
$Ru(bpm)_3^{2+}$	450	635	131	+1.69	−0.91	24
$Ru(bpy)_2bpm^{2+}$	480	710[d]	76[d]	+1.40	−1.02	83
$(bpy)_2Ru(bpm)Ru(bpy)_2^{4+}$	606[e]			+1.53	−0.41	83
$Ru[bpm)Ru(bpy)_2]_3^{8+}$	613[e]			[f]		83
$Ru(dpp)_3^{2+}$	455[e]	636[e]	270[e]	+1.68	−0.95	24
$Ru(bpy)_2dpp^{2+}$	464	660	226	+1.33	−1.06	83
$Ru(phen)_2dpp^{2+}$	465	652	252	+1.39	−1.07	83
$Ru(tpy)(dpp)Cl^{+}$	514			+0.9	−1.24	83
$(bpy)_2Ru(dpp)Ru(bpy)_2^{4+}$	523	756	134	+1.38	−0.66	83
$(phen)_2Ru(dpp)Ru(phen)_2^{4+}$	525	746	153	+1.44	−0.64	83
$(phen)_2Ru(dpp)Ru(bpy)_2^{4+}$	523	752	113			83
$(tpy)ClRu(dpp)RuCl(tpy)^{2+}$	561			+0.91	−0.82	83
$(bpy)Ru[(dpp)Ru(bpy)_2]_2^{6+}$	545	766	80	+1.48[g]	−0.71[h]	96
$(bpy)Ru[2,5\text{-}dpp)Ru(bpy)_2]_2^{6+}$	592	814	64	+1.45[g]	−0.66[h]	96
$(bpy)Ru[(dpp)Ru(biq)_2]_2^{6+}$	546	742	142	+1.62[g]	−0.63[h]	96
$(bpy)Ru[(2,5\text{-}dpp)Ru(biq)_2]_2^{6+}$	591	774	120	+1.62[g]	−0.55[h]	96
$Ru[(dpp)Ru(bpy)_2]_3^{8+}$	534	772	89	+1.50	−0.56	98
$Ru[(dpp)Ru(phen)_2]_3^{8+}$	539	760	87	+1.43	−0.50	98
$Ru[(dpp)Ru(tpy)Cl]_3^{5+}$	564	758	84	+1.06	−0.60	98
$Ru(dpq)_3^{2+}$	500	716	82	+1.70	−0.60	83, 99
$Ru(bpy)_2dpq^{2+}$	517	766	71	+1.42	−0.77	83
$Ru(phen)_2dpq^{2+}$	516	756	83	+1.42	−0.79	83
$(bpy)_2Ru(dpq)Ru(bpy)_2^{4+}$	603	822	<20	+1.47	−0.37	83
$(phen)_2Ru(dpq)Ru(phen)_2^{4+}$	601	810	<20	+1.48	−0.40	83
$(phen)_2Ru(dpq)Ru(bpy)_2^{4+}$	603	820	<20			83
$(bpy)Ru[(dpq)Ru(bpy)_2]_2^{6+}$	621			+1.57	−0.29	99
$Ru[(dpq)Ru(bpy)_2]_3^{8+}$	618			+1.60	−0.17	99

$Ru(tpp)_2^{2+}$	476	658	80			83
$Ru(tpy)(tpp)^{2+}$	472	684	91	+1.40	−0.94	83
$(tpy)Ru(tpp)Ru(tpy)^{4+}$	650	826	100	+1.43	−0.34	83
$Ru[(tpp)Ru(tpy)]_2^{6+}$	720			+1.04	−0.50	83
$Ru(bpy)_2(ppz)^{2+}$	474	700	200	+1.37	−1.11	81
$(bpy)_2Ru(ppz)Ru(bpy)_2^{4+}$	573	820	< 50	+1.35	−0.67	81
$Ru(bpy)_2(tppq)^{2+}$	573			+1.42	−0.42	100
$(bpy)_2Ru(tppq)Ru(bpy)_2^{4+}$	642			+1.45	−0.16	100
$Ru(bpy)_2(bidpq)^{2+}$	525	770	< 60	+1.41	−0.72	94
$(bpy)_2Ru(bidpq)Ru(bpy)_2^{4+}$	528	780	< 60	+1.42	−0.67	94
$[Ru(bpy)_2]_3(bidpq)^{6+}$	616			+1.46	−0.31	94
$[Ru(bpy)_2]_4(bidpq)^{8+}$	622			+1.46	−0.22	94
$Ru(bpy)_2HAT^{2+}$	432, 484sh[e]	745[e]	105[e]	+1.56	−0.84	92
$(bpy)_2Ru(HAT)Ru(bpy)_2^{4+}$	572[e]	825[e]	148[e]	+1.53	−0.49	92
$[Ru(bpy)_2]_3HAT^{6+}$	580[e]	880[e]	40[e]	+1.61	−0.25	92
$Ru(bpy)_2(bpt)^{+}$	480	678	160	+0.85	−1.47	87
$(bpy)_2Ru(bpt)Ru(bpy)_2^{3+}$	453	648	100	+1.04	−1.40	87

[a] In room temperature deaerated CH_3CN, solutions unless otherwise noted; [b] low-energy MBCT band; [c] potentials vs SCE; [d] propylene carbonate; [e] water; [f] irreversible process; [g] bielectronic process; [h] average value of two overlapping waves.

trinuclear complex $Ru[(tpp)Ru(tpy))]_2^{6+}$ [83]. This complex, which is nonluminescent in fluid solution, displays an unexpected red shift in absorption and a less positive oxidation potential with respect to that of the corresponding binuclear complex [83]. No explanation for this behavior was given.

The complexes $(bpy)_2Ru(bpm)Ru(bpy)_2^{4+}$ and $(L)_2Ru(dpp)Ru(L)_2^{4+}$ (L = bpy, phen, dc-bpy^{2-}) have been recently studied by Kalyanasundaram and Nazeeruddin [84], with the aim to clarify the origin of the wide differences in the behavior of bpm-bridged and dpp-bridged binuclear complexes. While the authors tend to agree with the previously reported observations on the dpp complexes, they report somewhat different results for the bpm complex. In particular the binuclear bpm complex is found to exhibit a previously unnoticed, extremely weak emission around 795 nm in room temperature aqueous solution. The authors conclude that the bpm- and dpp-bridged complexes are less different than previously believed. The fact still remains, however, that the effect of binucleation on luminescence (reflected by the red shift and the decrease in intensity of the emission in going from mononuclear to binuclear species) is much more pronounced for the bpm than dpp species.

The binuclear complex, $Ru(bpy)_2(bpt)Ru(bpy)_2^{3+}$, containing the bpt^- bridging ligand, has been studied by Hage et al. [85, 86] and Barigelletti et al. [87] together with the corresponding mononuclear $Ru(bpy)_2(bpt)^+$ species. In this binuclear complex, both emission and photochemistry (release of the $Ru(bpy)_2^{2+}$ fragment) were observed. Contrary to what happens with the other polynuclear complexes discussed in this section, in these systems the bpt^- bridge is more difficult to reduce than bpy so that the first reduction occurs at the bpy ligands and the Ru → bpy MLCT state is the lowest excited state. In going from the mononuclear to the binuclear complex an anodic shift in the oxidation potential and a parallel blue shift of the Ru → bpy MLCT transitions in the absorption as well as in the emission spectrum are observed. These results are consistent with a decrease of electronic charge on the Ru(II) ion caused by binucleation. An interesting feature of the binuclear complex is that, because of the non-equivalence of the 1 (or 2) and 4 nitrogens of the triazole ring in the bpt^- bridging ligand, the energy levels of the two $Ru(bpy)_2^{2+}$ fragments are not isoenergetic [87]. In particular, it is found that in this complex the lowest Ru → bpy MLCT state responsible for the luminescence is centered on the $Ru(bpy)_2^{2+}$ unit bound to the 1 (or 2) position, whereas photochemistry takes place from a ligand field LF state centered on the $Ru(bpy)_2^{2+}$ unit bound to the 4 positions. No definite evidence for the occurrence of intramolecular energy transfer has been obtained in this system.

3.2.2 Heterometallic Binuclear Polyimine Complexes

The relevant properties of the complexes discussed in this section are reported in Table 3.

The $(bpy)_2Ru(bpm)PtCl_2^{2+}$ complex has been studied by Sahai and Rillema [88]. The absorption spectrum of the complex is quite similar in the visible region to that of the parent homobinuclear ruthenium complex, $(bpy)_2Ru(bpm)Ru(bpy)_2^{4+}$ except for a slight blue shift of the low energy MBCT band. On the other hand, in contrast to the homobinuclear Ru complex, the electrochemical behavior indicates little communication between the metal centers. On the basis of previous arguments about the metal-metal communication, this complex would be expected to be luminescent. No information about the emissive properties of this heterobinuclear complex, however, is reported.

Table 3. Spectroscopic, Photophysical and Redox Properties of Homo- and Hetero Polynuclear Polyimine Complexes[a]

Complexes	Absorption[b] λ_{max}, nm	Emission λ_{max}, nm (τ, ns)		Redox Properties[c] $E^{ox}_{1/2}$, V	$E^{red}_{1/2}$, V	Ref.
$Os(dpp)_3^{2+}$	475	768	82	+0.92	−0.84	97
$Os(bpy)_2(dpp)^{2+}$	486[d]	776	60	+0.85[d]	−1.05[d]	89
$Ru(bpy)_2dpp^{2+}$	464	660	226	+1.33	−1.06	83
$(bpy)_2Os(dpp)Os(bpy)_2^{4+}$	550[d]			+0.83	−0.72	89
$(bpy)_2Os(dpp)Ru(bpy)_2^{4+}$	534[d]			+0.90	−0.73	89
$Os[(dpp)Ru(bpy)_2]_3^{8+}$	549	875	18	+1.25	−0.55	97
$(phen)_2Ru(dpp)Fe(CN)_4$	476	700	90			83
$(phen)_2Ru(dpp)Fe(CN)_4^+$	498	680	95			83
$(bpy)_2Os(bpm)Ru(bpy)_2^{4+}$	515					89
$(bpy)_2Ru(bpm)PtCl_2^{2+}$	571[e]			+1.35[e]	−0.34[e]	88
$Ru(bpq)_3(PtCl_2)_3^{2+}$	553[e]				−0.07[e]	95

[a] In room temperature, deaerated CH_3CN solutions unless otherwise noted; [b] low-energy MBCT band; [c] potentials vs SCE; [d] DMF; [e] propylene carbonate.

The $(bpy)_2Ru(dpp)Os(bpy)_2^{4+}$ complex has been studied by Kalyanasundaram and Nazeeruddin [89], together with the parent homobinuclear Os complex, $(bpy)_2Os(dpp)Os(bpy)_2^{4+}$. No emission was observed up to 850 nm for these systems. The authors point out that, on the basis of the absorption spectrum, the emission is expected to lie beyond the detection limit of the equipment. The shifts in absorption and redox properties caused by binucleation in these dpp-bridged complexes are larger than in the analogous 4,4′-bpy- and bpa-bridged homobinuclear complexes (Sect. 3.3.4). On the basis of these results, these binuclear complexes are considered by the authors as examples of "strongly coupled" cases. It should be noticed, however, that the similarity of the first oxidation potential with that of the mononuclear Os analogue would be, according to Petersen [83], an indication of weak metal-metal electronic interaction.

Preliminar results on the $(phen)_2Ru(dpp)Fe(CN)_4$ and $(phen)_2Ru(dpp)Fe(CN)_4^+$ complexes have been reported by Petersen [83]. Both these binuclear complexes are found to emit in room temperature fluid solution. This result is surprising, as the Fe-based fragments are expected to quench the emission of the ruthenium chromophore on the basis of the well known behavior of analogous Fe(II) and Fe(III) complexes in bimolecular processes [90, 91]. A tentative explanation of this puzzling result based on the dpp bridge structure is discussed by the author [83].

3.2.3 Polyimine Complexes of Higher Nuclearity

The relevant properties of the homometallic and heterometallic complexes discussed in this section are reported in Tables 2 and 3, respectively.

Ru(II) polynuclear complexes containing more than two metal sites have been built up following two strategies: i) by attaching to a common tris- or tetra-chelating ligand three or four metal containing units; ii) by attaching peripheral metal-containing units to a $M(B)_3^{2+}$ core. Examples of these two types of aggregations are given below.

In the trimetallic $[Ru(bpy)_2]_3HAT^{6+}$ complex studied by Masschelein, et al. [92] and Kirsh-De Mesmaeker et al. [93], three $Ru(bpy)_2^{2+}$ units are bound around the central symmetrical HAT ligand. The most interesting result is that this complex, as well as the corresponding mononuclear and binuclear species, exhibits emission in room temperature fluid solution from a Ru → HAT charge transfer state [92]. This, together with the similarity in the first oxidation potential in the series mono-, bi-, trinuclear species, suggests that, in spite of the planar structure of the HAT bridge, substantial communication between the metal centers across the bridge does not occur. Similar results have been reported by Gafney [81] for the planar ppz bridged binuclear complex (Sect. 3.2.1).

Similar aggregates where one, two, three or four $Ru(bpy)_2^{2+}$ equivalent units are bound to a single bidpq ligand have been studied by Rillema and coworkers [94]. In this series, the mono and binuclear species are found to emit very weakly, while no luminescence has been observed for the trinuclear and tetranuclear complexes. Spectroscopic and electrochemical data for this series of complexes suggest that the bidpq bridging ligand can be described as containing two independent dpq subunits. This explains why the binuclear complex, in which the two Ru centers are coordinated to different dpq-type subunits of the bridging ligand, is an emitting species. With the same bidpq multidentate bridge, Sahai and Rillema [95] have synthesized the hetero tetranuclear $Ru(bidpq)(PtCl_2)_3^{2+}$. No emission data are reported for this complex.

Trinuclear complexes of general formula $(L')Ru[BRu(L)_2]_2^{6+}$, where L = bpy or biq, L′ = bpy, and B = dpp or 2,5-dpp, have been studied by Balzani and coworkers [96]. As the parent binuclear dpp complex discussed in Sect. 3.2.1, all the complexes are luminescent in room temperature fluid solution. These complexes are interesting since they contain two non-equivalent Ru sites: one central (Ru_c) and two peripheral (Ru_p) metal centers (Eq. 23)

$$(L)_2Ru_p(B)Ru_c(bpy)(B)Ru_p(L)_2 \tag{23}$$

For each complex at least four different types of metal-to-ligand charge transfer transitions are expected: Ru_c → bpy, Ru_c → B, Ru_p → B and Ru_p → L. Two broad bands are observed for each complex in the visible region. The low-energy band receives contributions mainly from CT transitions involving the bridging ligand. The bielectronic nature of the first oxidation wave indicates that oxidation first occurs at the two peripheral Ru centers. This result suggests that the Ru_p → B is the lowest excited state responsible for the emission. It seems likely that, as suggested by the same authors for the analogous tetrametallic complex discussed below, intramolecular energy tranfer process from upper MLCT states to the $Ru_p \rightarrow B$ CT state occurs in this system.

A fairly interesting case of larger aggregate is the hetero-tetranuclear $Os[(dpp)Ru(bpy)_2]_3^{8+}$ studied by Balzani and coworkers [97]. The complex consists of a central $Os(dpp)_3^{2+}$ core with three bound $-Ru(bpy)_2^{2+}$ peripheral units. Interestingly, this complex exhibits emission in room temperature fluid solution from a triplet Os → dpp MBCT state. This assignment is consistent with the electrochemical behavior: reduction first takes place at the dpp ligand and oxidation first occurs at the central osmium core. The important observation is that the luminescence from the Os → dpp CT state is obtained with the same efficiency regardless of the Ru → bpy, Ru → dpp or Os → dpp type of excitation. This result indicates that intramolecular energy transfer from the peripheral Ru-containing units to the central Os-containing core occurs with unitary efficiency. As far as the mechanism is concerned, although a Coulombic singlet-singlet energy transfer cannot be excluded, an exchange triplet-triplet mechanism seems to be more likely [97].

In the analogous homo-tetrametallic complex where the central Os^{2+} ion is replaced by Ru^{2+}, $Ru[(dpp)Ru(bpy)_2]_3^{8+}$, studied first by Petersen and coworkers [98] and reinvestigated by Balzani and coworkers [97], the energy levels of the central unit are higher than the corresponding levels of the peripheral units. This is consistent with the fact that the central Ru site bound to three dpp ligands, that are stronger π-acceptors than bpy, is more difficult to oxidize than the peripheral Ru sites (bound to one dpp and two bpy ligands). In this system, the intramolecular energy transfer process occurs from the central unit to the peripheral ones. Both these tetranuclear complexes, as pointed out by Balzani and coworkers [97], feature the so-called "antenna effect", but in opposite directions: in the heterometallic complex luminescence takes places from the central chromophore, which receives the energy collected by the peripheral chromophores, whereas the reverse occurs in the homotetranuclear complex.

Homo-tetrametallic complexes structurally similar to the above-discussed complex, buth with bpm and dpq bridging ligands instead of dpp, have been studied by Ludi et al. [78] and by Rillema et al. [99]. For the tetranuclear complex with the strongly coupling bpm bridge, Ludi reported that no luminescence could be detected up to 900 nm in solid matrix at 15 K, while the analogous binuclear complex exhibits an emission around 800 nm in the same conditions. For the bpq tetranuclear complex, as for its binuclear analogue [100], no luminescence data were reported by Rillema et al. [99].

The heptanuclear $Ru(bpz)_3[Ru(NH_3)_5]_6^{14+}$ represents a remarkable example of a very weak coupling case among the polyimine bridged polynuclear complexes of this type. This complex has been studied by Lever and coworkers [101], who have taken advantage of the fact that the $Ru(bpz)_3^{2+}$ has six peripheral uncoordinated nitrogen atoms to build up the whole series of di- to heptanuclear complexes with $Ru(NH_3)_5^{2+}$ units. The spectra of the $Ru(bpz)_3[Ru(NH_3)_5]_n^{(2+2n)+}$ (n = 1, 3, 6) complexes, in addition to the MLCT Ru → bpz bands present in the $Ru(bpz)_3^{2+}$ monomer, exhibit two additional bands at lower energy. These bands, whose intensity increases with the number of the $Ru(NH_3)_5^{2+}$ groups, are assigned to charge transfer transitions from the peripheral Ru centers to the bridge ($Ru_p \rightarrow$ bpz MBCT). The following important observations are reported: i) the energies of the MBCT $Ru_p \rightarrow$ bpz bands and the electrochemical potentials for the $Ru_p(III)/Ru_p(II)$ couples vary only slightly with the number of $Ru(NH_3)_5^{2+}$ groups, ii) the MLCT transitions from the central Ru to bpz are completely unaffected by peripheral coordination, even when all the Ru_p centers are oxidized from the 2+ to the 3+ oxidation state (mixed-valence species), iii) no evidence for IT bands with $\varepsilon > 100\ M^{-1}\ cm^{-1}$ has been detected in the near IR region of the mixed-valence species. These results indicate that the peripheral Ru_p centers are essentially uncoupled to each other and to the central ruthenium atom. The authors have noted that the situation of this polynuclear complex is completely different from that of analogous systems with a linear structure studied by Taube and coworkers [102], where extended interactions are observed. Although no data concerning the photophysical behavior are reported, in the polynuclear species the MLCT emission of the $Ru(bpz)_3^{2+}$ central unit could likely be quenched by the peripheral $Ru(NH_3)_5^{2+/3+}$ fragments via intramolecular energy and/or electron transfer process (see Sects. 3.3.3, 3.3.6, and 3.3.7 for related systems).

A similar heptanuclear complex consisting of the $Ru(bpz)_3^{2+}$ central unit with bound pentacyanoferrate(II) units has been reported by the same authors [103]. Similar results pointing towards very weak electronic interaction were obtained also in this case.

3.2.4 Polyimine-Bridged Carbonyl Complexes

This section brings together a number of studies in which the following categories of complexes have been characterized: a) homo- and hetero-binuclear $(CO)_5M(B)M'(CO)_5$ complexes, where M, M' = Mo, Cr, or W; b) homobinuclear $Cl_2(CO)_2M(B)M(CO)_2Cl_2$ complexes, where M = Ru, and $Cl(CO)_3M(B)M(CO)_3Cl$ complexes, where M = Re; c) hetero binuclear and polynuclear complexes of the type $[(CO)_3ClM(B)]_{3-n}M'(bpy)_n^{2+}$, where n = 0–2, M = Re, M' = Ru. The spectroscopic, photophysical and redox properties of the polynuclear complexes examined here are listed in Table 4 together with the properties of the corresponding mononuclear species.

The homobinuclear $(CO)_5W(B)W(CO)_5$ complexes with B = 4,4'-bpy, pz, bpe and bpa and heterobinuclear $(CO)_5M(4,4'\text{-bpy})M'(CO)_5$ complexes with M, M' = = Cr, Mo, W have been studied by Lees and coworkers [104], together with the corresponding mononuclear species. Each complex exhibits low-lying ligand field (LF) and MBCT absorptions. The energy position of the latter is dependent on both the length and conjugation of the bridge. When B = pz, 4,4'-bpy, or bpe, the MBCT state is the lowest excited state, while when B = bpa the LF states are the lowest-lying states both in the mononuclear and binuclear complexes. There is no effect on the LF transitions upon binucleation. On the other hand, different shifts in the MBCT transitions with respect to the corresponding mononuclear species are observed, depending on the type of bridge: in the 4,4'-bpy and

Table 4. Spectroscopic, Photophysical and Redox Properties of Mono- and Polynuclear Polyimine Carbonyl Complexes[a]

Complexes	Absorption[b] λ_{max}, nm	Emission λ_{max}, nm	(τ, ns)	Redox properties[c] $E^{ox}_{1/2}$, V	$E^{red}_{1/2}$, V	Ref.
$(CO)_5W(pz)$	[d]	645	18	+1.25[e,f]	−1.70[e,f]	104
$(CO)_5W(pz)W(CO)_5$	510	721	186	+1.26[e,f]	−1.21[f]	104
$(CO)_5W$(4,4′-bpy)	[d]	643	415	+1.09[e,f]	−1.64[e,f]	104
$(CO)_5W$(4,4′-bpy)$W(CO)_5$	438	651	428	+1.09[e,f]	−1.30[f]	104
$(CO)_5W$(bpe)	440sh	549	120	+1.13[e,f]	−1.50[e,f]	104
$(CO)_5W$(bpe)$W(CO)_5$	450	560	123	+1.09[e,f]	−1.20[f]	104
$(CO)_5W$(bpa)	340sh					104
$(CO)_5W$(bpa)$W(CO)_5$	343			+1.16[e,f]	> −2.3[f]	104
$(CO)_5W$(4,4′-bpy)$Cr(CO)_5$	436sh	642	368	+1.06[e,f]	−1.37[f]	104
$(CO)_5Mo$(4,4′-bpy)	428sh	595	699	+1.16[e,f]	−1.40[e,f]	104
$(CO)_5W$(4,4′-bpy)$Mo(CO)_5$	432sh	651	428	+1.07[e,f]	−1.32[f]	104
$(CO)_5Mo$(4,4′-bpy)$Mo(CO)_5$	430sh	606	547	+1.16[e,f]	−1.38[f]	104
$(CO)_5Mo$(4,4′-bpy)$Cr(CO)_5$	432sh	602	767	+1.06[e,f]	−1.43[f]	104
$Re(CO)_3Cl$(bpm)	371[g]	567[h]		+1.43[e,i]	−1.03[i]	108, 110
$Cl(CO)_3Re$(bpm)$Re(CO)_3Cl$	465[i]			+1.48[j]	−0.51[j]	109
$(bpy)_2Ru$(bpm)$Re(CO)_3Cl^{2+}$	556[g]	774[k]		+1.76[i]	−0.41[i]	108, 110
(bpy)Ru[(bpm)$Re(CO)_3Cl]_2^{2+}$	531[i]	630[i]	942[i]	+1.61[i]	−0.33[i]	110
Ru[(bpm)$Re(CO)_3Cl]_3^{2+}$	501[i]	640[i]	847[i]	+1.59[i]	−0.21[i]	110
$Br(CO)_3Re$(QP)$Re(CO)_3Br$	370[j]	583[j]	960[j]	+0.98[j]	−1.45[j]	109
Ru(dpp)$(CO)_2Cl_2$	385			+1.90[e,i]	−1.04[i]	106
$Cl_2(CO)_2Ru$(dpp)$Ru(CO)_2Cl_2$	410			+2.00[e,i]	−0.62[i]	106
Ru(2,5-dpp)$(CO)_2Cl_2$	405			+1.86[e,i]	−0.93[i]	106
$(CO)_2Cl_2(CO)_2Ru$(2,5-dpp)$(CO)_2Cl_2$	448[i]			+1.92[e,i]	−0.38[i]	106
$(bpy)_2Ru$(HAT)$[Re(CO)_3Cl]_2^{2+}$	529[i]	664[k]		+1.74[i]	−0.07[i]	110

[a] In room temperature, deaerated, benzene solutions unless otherwise noted; [b] low-energy MBCT band; [c] potentials vs SCE; [d] overlaps substantially with LF transitions; [e] irreversible process; [f] methylene chloride; [g] DMSO; [h] solid sample; [i] CH_3CN; [j] DMF; [k] 77 K.

bpe cases the spectra of the mononuclear and binuclear complexes are fairly similar, whereas in the pz case a substantial red shift of the MBCT transition was observed. These spectral shifts are reflected in parallel shifts in the potentials for reduction of the bridging ligand. The potentials for oxidation of the metal centers are not significantly different from mononuclear to binuclear species, suggesting a relatively weak metal-metal interaction. As far as the emission properties are concerned, when B = bpa no emission was observed at room temperature from either the mononuclear or binuclear complex, but an emission assigned to LF state was observed at 80 K. When B = 4,4′-bpy, bpe or pz, both mononuclear and binuclear complexes exhibit a broad MBCT emission in room temperature fluid solution as well as at 77 K. For some of the binuclear complexes, this broad MBCT emission was found to be dual. The origin of this phenomenon has been discussed by Lees and coworkers [105]. This series of complexes represents an example in which, depending on the nature of the bridge, the properties of mononuclear complexes are essentially unperturbed (B = bpa), weakly perturbed (B = 4,4′-bpy, bpe), and strongly perturbed (B = pz) on forming the binuclear complexes.

Binuclear complexes of the type $Cl_2(CO)_2Ru(B)Ru(CO)_2Cl_2$ with B = dpp or 2,5-dpp have been studied by Balzani and coworkers [106], together with the corresponding mononuclear species. Besides ligand centered bands in the UV region, the binuclear complexes exhibit a broad MBCT absorption in the visible region, red shifted with respect to the corresponding mononuclear derivatives. All the binuclear complexes, as well as the corresponding monomeric species, exhibit a long-lived ligand-centered (LC) emission at 77 K but are non-luminescent in room-temperature fluid solution, where they are strongly photosensitive. These results have been interpreted [106] on the basis of a temperature-dependent interplay of closely lying LC, MBCT, LF. On binucleation, the MBCT states move down in energy while the LF states are practically unaffected. This causes a decrease in the photoreactivity in going from mononuclear to binuclear species. As far as the redox behavior is concerned, no significant change in the first oxidation potential was observed indicating that little metal-metal interaction occurs through the non-planar dpp bridge. This result is in full agreement with the behavior of other binuclear dpp-bridged carbonyl complexes [107] and also of other dpp-bridged Ru polynuclear complexes (Sects. 3.2.1, 3.2.2 and 3.2.3).

The $Cl(CO)_3Re(bpm)Re(CO)_3Cl$ complex was investigated first by Vogler and Kisslinger [108]. This complex was found to be nonemissive even in the solid state at 4 K, in sharp contrast to the highly emissive nature of the corresponding monometallic species. This nonemitting behavior was attributed by the authors to the presence of a low-lying nonemissive metal-to-metal charge transfer (IT) excited state for which, however, no evidence was found in the absorption spectrum. More recently Juris et al. [109] reexamined this bpm-bridged complex together with the analogous QP-bridged species. They reported that, while the $Cl(CO)_3Re(QP)Re(CO)_3Cl$ complex was found to exhibit a MBCT emission in fluid solution, no detectable emission was observed for the $Cl(CO)_3Re(bpm)$-$Re(CO)_3Cl$ in agreement with the result reported by Vogler. The authors discuss this difference in behavior on the basis of metal-metal communication arguments.

In the bpm-bridged binuclear complex, as well as in the analogous polynuclear complexes of Ru (see Sect. 3.2.1 and 3.2.2), the bpm bridge permits communication between the two metals, so that emission is not observed, whereas the QP ligand hinders communication between the two Re centers in the emissive $Cl(CO)_3Re(QP)$-$Re(CO)_3Cl$ complex.

The $(bpy)_2Ru(bpm)Re(CO)_3Cl^{2+}$ complex has been studied by Vogler and Kisslinger [108]. Interestingly, this cation was found to emit at 77 K from the lowest-energy Ru → bpm MBCT state regardless of the Re → bpm or Ru → bpm nature of the MBCT bands irradiated. This behavior is interpreted by the authors in terms of intramolecular "energy transfer" from the Re-containing fragment to the emissive MBCT state of the Ru-containing fragment (Eq. 24).

$$(bpy)_2Ru^{II}(bpm^{-\cdot})Re^{II}(CO)_3Cl^{2+} \rightarrow (bpy)_2Ru^{III}(bpm^{-\cdot})Re^{I}(CO)_3Cl^{2+} \quad (24)$$

Actually, as pointed out by Vogler and Kisslinger [108], this process can be more precisely described as a Ru(II) → Re(II) electron transfer process.

Rillema and coworkers [110] have extended the study of Vogler and Kisslinger by including the trinuclear complexes, $(bpy)Ru[(bpm)Re(CO)_3Cl]_2^{2+}$, and $(bpy)Ru[(HAT)Re(CO)_3Cl]_2^{2+}$ and the tetranuclear complex $Ru[(bpm)Re(CO)_3Cl]_3^{2+}$. An intriguing result of this study is that while the trinuclear HAT-bridged complex exhibits emission only at 77 K the trinuclear and tetranuclear bpm-bridged complexes were found to emit at room temperature, though very weakly, with a relatively long lifetime. The reason of this result, which is in sharp contrast to the general nonemitting behavior observed for bpm-bridged complexes of this type, is not discussed by the authors. As far as the origin of the emission is concerned, the authors tentatively assign it as a MBCT emission of the Ru-containing fragment, based on the relatively long lifetime. The authors discuss the properties of this series of complexes in terms of weak electronic interactions between the metal centers [110].

3.2.5 Remarks

In the papers surveyed in this section (3.2.1–3.2.4), the main emphasis is placed on the determination of the degree of electronic interaction between the metal-containing fragments and on its relationship with the emissive properties of the polynuclear complexes. In this regard, Gafney's concept [80] that weakly coupled systems are emissive whereas strongly coupled systems are not has often been used as a general framework to discuss experimental results. From the number of apparent exceptions and ambiguous cases noticed in the previous survey, however, further work appears to be needed to arrive at a consistent picture of the photophysics of these polynuclear complexes.

The discussion of the behavior of these polynuclear complexes is complicated by the fact that different authors seem to use the terms "strong" and "weak" coupling in somewhat different ways. Basically, four types of independent experimental approaches have been used to estimate the degree of interaction between metal-containing subunits in these polynuclear complexes: (i) differences in the first potential for reduction of the bridging ligand between mononuclear and polynuclear species; (ii) spectral shifts in MBCT transitions between mononuclear and polynuclear complexes; (iii) differences in the first potential for oxidation of the metals between mononuclear and polynuclear complexes; (iv) differences between the first and second potential for oxidation of the metals in a polynuclear complex (comproportionation constants). These methods reflect the stabilization of the bridging ligand π^* orbitals (methods (i) and (ii)) and metal *d* orbitals (methods (iii) and (iv)) upon coordination of the second metal to the bridging ligand. The stabilization of bridging ligand π^* orbitals is a first-order effect that is expected to occur, to a greater or lesser extent, for virtually all of the bridging ligands dealt with in this section (with the exception of the "insulating" bpa bridge). On the other hand, stabilization of metal *d* orbitals is a weaker, second-order effect that may become sizeable only for truly delocalizing bridges. Thus, the terms "weak" and "strong" coupling should be used in a relative sense, depending on the systems being compared and on the parameters chosen to estimate the interaction. It appears that most of the authors tend to use "weak coupling" for systems, such as e.g. the 4,4′-bpy, bpe, dpp bridged complexes, in which effects (i) and (ii) are sizeable but (iii) and (iv) are not, and "strong coupling" for the systems, such as e.g. bpm bridged complexes, in which effects (iii) and (iv) are evident. It should be noticed that, in this sense, the polynuclear complexes discussed in section 3.1 should be considered as completely "uncoupled" systems, since in those cases not even effects (i) and (ii) were observed.

A point that remains somewhat obscure after the literature survey is the *mechanism* that underlies the gross correlation between metal-metal coupling and emission properties. For typical "weakly coupled" cases, it appears that the emission properties can generally be accomodated within conventional models of MLCT excited-state behavior [24], simply by considering the effects (energy ordering vs d—d states, energy-gap law) of the decrease in energy of the MBCT state induced by binucleation. What is not clear is whether, for typical "strongly coupled" cases, this model is still appropriate (allowing for the larger energy change and lifetime shortening) or if, on the contrary, a qualitatively new model of delocalized molecular states of the polynuclear complex is implied. Some authors allude to this problem in terms that, though somewhat vague, seem to favor the second viewpoint. According to Petersen [98], the second metal center may act by "coupling of other excited states and deactivation processes into the manifold of the first metal center". According to Gafney [80], "the π^* acceptor orbital may be distorted toward the second metal center which, if coupled to the solvent, may offer an efficient pathway for radiationless deactivation". Clearly, a definite picture of this effect has not yet been established.

3.3 Energy and Electron Transfer Pathways in the Deactivation of Ruthenium, Rhenium and Osmium Polynuclear Complexes

This section brings together a number of studies in which particular emphasis is placed on the availability of multiple pathways for excited state deactivation in polynuclear complexes.

Schematically, the binuclear complexes dealt with in this section have a structure of the type

$$L_a{-}M_a{-}B{-}M_b{-}L_b$$

where M_a and M_b represent metal centers, L_a and L_b stand for the nonbridging ligands in their coordination spheres, and B is a bridging ligand. In the complexes discussed in this section, L_a and/or L_b constitute relatively low-energy redox sites, and thus can be involved in low-energy charge transfer transitions. In this sense, L_a and/or L_b can be considered "chromophoric" ligands. The B bridging ligand may often also constitute a low-energy redox site, and thus be a "chromophoric" ligand too. In general, the M_a and M_b units in these complexes can be easily oxidized. On the other hand they usually lack low-lying metal-centered states. In multi-site systems of this type, a variety of metal-to-ligand charge transfer excited states is available, as shown by Scheme I, where state (Ia) is the ground state, states (Ib) and (Ic) are metal-to-nonbridging ligand charge transfer states (in this context referred to as MLCT states), states (Id) and (Ie) are metal-to-bridging ligand charge transfer states (in this context referred to as MBCT states), and states (If) and (Ig) are "remote" MLCT excited states in which the metal and the ligand involved in the transition are not directly bound to each other.

$$L_a{-}M_a{-}B{-}M_b{-}L_b \quad \text{(Ia)}$$
$$L_a^-{-}M_a^+{-}B{-}M_b{-}L_b \quad \text{(Ib)}$$
$$L_a{-}M_a{-}B{-}M_b^+{-}L_b^- \quad \text{(Ic)}$$
$$L_a{-}M_a^+{-}B^-{-}M_b{-}L_b \quad \text{(Id)}$$
$$L_a{-}M_a{-}B^-{-}M_b^+{-}L_b \quad \text{(Ie)}$$
$$L_a^-{-}M_a{-}B{-}M_b^+{-}L_b \quad \text{(If)}$$
$$L_a{-}M_a^+{-}B{-}M_b{-}L_b^- \quad \text{(Ig)}$$

Scheme I

This leads to the prediction of a rich charge transfer spectroscopy for these species, with intriguing facets related to the observability of remote MLCT transitions. On the other hand, the radiationless processes interconverting these excited states of the binuclear complex can be seen as intramolecular *electron* or *energy* transfer processes depending on whether one electron or two electrons are exchanged between the sites. Thus, e.g. (Ib) ↔ (Ic) is an energy transfer, (Ib) ↔ (Id) is an electron transfer (of the L ↔ B type), (Ib) ↔ (Ie) is an energy transfer, and (Ib) ↔ (If) is an electron transfer (of the $M_b \rightarrow M_a$ type). Direct deactivation of these charge transfer states to the Ia ground state, of course, can always be considered as an intramolecular electron transfer process. The above scheme holds for a binuclear complex but can be easily extended, allowing for the increase in complexity, to tri- and polynuclear complexes of the same type.

Since the metal units can be easily oxidized, it is in principle also possible to study the photophysical behavior of the singly oxidized form of these complexes (that of the doubly oxidized form is not interesting due to the disappearence of all charge transfer states upon two-electron oxidation). For a singly oxidized form such as

$$L_a{-}M_a^+{-}B{-}M_b{-}L_b$$

a variety of charge transfer excited states are also possible, as shown in Scheme II. State (II a) is the ground state, state (II b) is an excited state in which the thermodynamically less favoured site is oxidized. This metal-to-metal charge transfer state is commonly denoted as intervalence transfer (IT) state (Sect. 2.2). Analogously to the previous case, (II c) is a MBCT state, (II d) is a MLCT state, and (II e) is a remote MLCT state. In principle, the spectroscopy of such systems is thus even richer than that of the reduced species, as additional features arising from IT bands are expected. In such systems, all the radiationless processes interconverting the excited states and leading to the ground state are to be considered one-electron transfer steps. Here again, the extension from the binuclear case to cases of higher nuclearity can be easily done.

$$L_a{-}M_a^+{-}B{-}M_b{-}L_b \qquad \text{(II a)}$$

$$L_a{-}M_a{-}B{-}M_b^+{-}L_b \qquad \text{(II b)}$$

$$L_a{-}M_a^+{-}B^-{-}M_b^+{-}L_b \qquad \text{(II c)}$$

$$L_a{-}M_a^+{-}B{-}M_b^+{-}L_b^- \qquad \text{(II d)}$$

$$L_a^-{-}M_a^+{-}B{-}M_b^+{-}L_b \qquad \text{(II e)}$$

Scheme II

Schemes I and II illustrate the variety of charge transfer excited states available in a polynuclear complex in which both the metals and the ligands represent accessible redox sites. These states provide a variety of pathways for deactivation following light absorption, the actual behavior of any particular complex depending on the actual energy ordering of the excited states and on the kinetic factors that govern the competition between the intramolecular processes. The studies discussed below provide experimental examples of various energy and electron transfer deactivation pathways and give insight into the factors that determine the photophysical behavior. In most of the studies reviewed, complex mechanistic discussions are reported. In the following sections, only the main photophysical pathways will be indicated for each complex. The reader is referred to the original literature for details.

3.3.1 $(NH_3)_5Ru(pz)Ru(EDTA)^+$

The study of this complex by Creutz et al. [111] represents one of the pioneering studies in the area. The complex is a mixed-valence species corresponding to the electronic structure

$$(NH_3)_5Ru^{II}(pz)Ru^{III}(EDTA)^+$$

The main spectral features of this complex are a Ru(II) → pz MBCT band in the visible and a Ru(II) → Ru(III) IT band in the near infrared. Picosecond laser photolysis of the corresponding fully reduced Ru(II)—Ru(II) species or of the mononuclear $Ru(NH_3)_5(pz)^{2+}$ model compound gives rise to a transient bleaching of the ground-state absorption with lifetimes of the order of 0.1 ns, characteristic of the MBCT state. With the Ru(II)—Ru(III) mixed valence species, on the other hand, a much less pronounced bleaching is observed (about 1/10 of that of the previously mentioned complexes). This small bleaching decays with a lifetime of 0.08 ns. This result is interpreted by the authors in terms of the pathway shown in Scheme III. In this scheme, a ligand-to-metal electron transfer process causes prompt quenching of the MBCT state and population of the lowest IT excited state. The IT state (that is considered to be responsible for the small bleaching observed) decays then back to the ground state by a metal-to-metal intramolecular

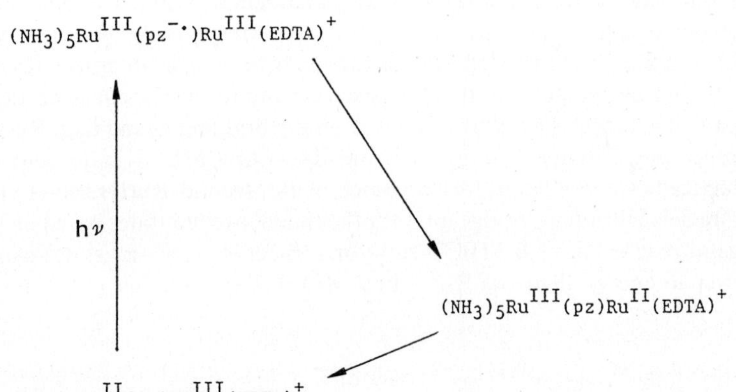

Scheme III

electron transfer process. The observed decay rate is in satisfactory agreement with the value calculated from redox potentials and spectroscopic IT band parameters on the basis of the Hush model (Sects. 2.1 and 2.2), provided that a frequency factor appropriate for solvent reorganizational modes is used [111]. A puzzling result, that does not easily fit into the mechanism of Scheme III, is the lack of any transient bleaching following direct excitation into the IT absorption band using near infrared laser pulses. A possible rationale is offered by the authors by assuming that, upon direct population by light absorption, the IT state may decay to the ground state prior to solvent relaxation [111].

A system bearing some similarity to the above one had been studied by Durante and Ford [112] a few years before. It consists of aqueous solutions in which $Ru(NH_3)_5(pz)^{2+}$ is associated via a pyrazine bridge to excess Cu^{2+} ions. Here too, excitation into the Ru(II) → pz MBCT absorption band gives rise to a transient bleaching that is attributed by the authors to formation of a Ru(II) → Cu(II) IT state. Contrary to the previous case, no IT band is observed in the absorption spectrum of this complex. The very slow recovery of the bleached absorption ($k \approx 8 \times 10^3\ s^{-1}$) is attributed to back Cu(I)-to-Ru(III) intramolecular electron transfer, although the possibility of a bimolecular process following dissociation of Cu^+ cannot be completely ruled out.

3.3.2 $(dpte)_2ClRu(B)Ru(bpy)_2Cl^{2+/3+}$ (B = bpa, bpe, 4,4′-bpy)

These binuclear systems were studied by Curtis et al. [113]. The 2+ and 3+ ions behave quite differently and will thus be discussed separately.

$$(dpte)_2ClRu^{II}(B)Ru^{II}(bpy)_2Cl^{2+}$$

The visible spectra are dominated by Ru → bpy MLCT absorption bands. With all bridging ligands, these complexes exhibit an emission that matches closely that of the mononuclear $Ru(bpy)_2Cl(B)^+$ analogues. With B = bpa and 4,4′-bpy, the emission lifetime is the same as for the corresponding mononuclear complex, but with B = bpe an appreciable reduction in lifetime is observed. The explanation proposed by the authors [113] is based on the following considerations: (i) with bpa and 4,4′-bpy, excited states based on the bridging ligand (e.g. Ru → B MBCT states), are at higher energy than the Ru → bpy MLCT state, and this state is practically unaffected by the presence of the second Ru(II) center; (ii) with the more delocalized bpe bridge, on the other hand, excited states based on the bridging ligand (e.g. a Ru → B MBCT state or a B-centered π—π* state) can be slightly lower in energy than the Ru → bpy MLCT state and can provide a reasonably

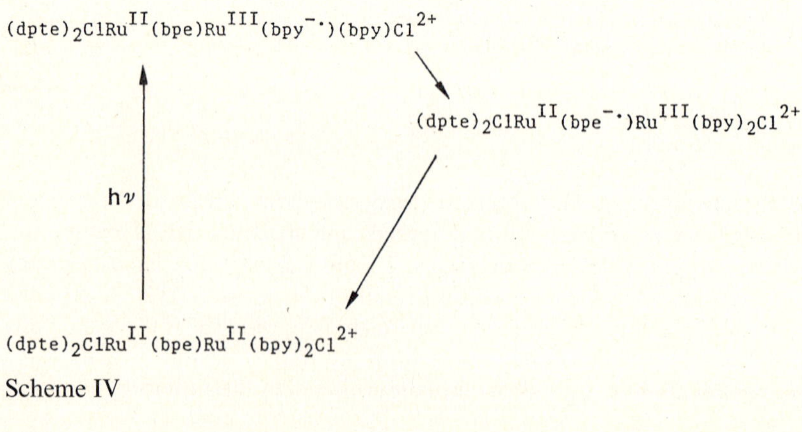

Scheme IV

efficient intramolecular quenching pathway. This last case is schematically depicted in Scheme IV. The key process converting the MLCT state to the MBCT one is an intramolecular ligand-to-ligand electron transfer process. The occurrence of a process of this kind implies that some electronic coupling is present between bpy and bpe in this complex. That ligand-ligand coupling can be sizeable in metal polypyridine-type complexes has been demonstrated by the observation of optical ligand-to-ligand charge transfer transitions in absorption and/or emission spectra [114–116].

Consistent with the redox properties of the two metal centers, the 3+ ion is a Ru(II)—Ru(III) complex of the type

$$(dpte)_2ClRu^{II}(B)Ru^{III}(bpy)_2Cl^{3+}$$

Visible excitation of the complex corresponds mainly to Ru(II) → B MBCT transitions (although some contribution by Cl → Ru(III) LMCT is also present). The interesting observation is that, with all B ligands, an emission characteristic of the $—Ru^{II}(bpy)_2—$ chromophore is present, although such a chromophore does not exist in the Ru(II)—Ru(III) complexes. The necessary conclusion [113] is that a B → bpy ligand-to-ligand electron transfer process occurs after excitation leading to the emitting state. This pathway is depicted in Scheme V. There are a number of interesting points concerning this scheme. First, with respect to the ground state, the emitting state can be viewed as a "remote" MLCT state involving metal and ligands that are not directly bound to each other. Furthermore, the state reached following the emission process is *not* the ground state but rather an IT state that must decay to the ground state by a metal-to-metal electron transfer process. This process cannot be detected in laser photolysis, presumably because of its occurrence in the subnanosecond time scale. On the other hand, the opposite optical process, i.e. IT absorption, is clearly seen in the near-infrared spectra of the complexes with B = bpe, 4,4′-bpy [113].

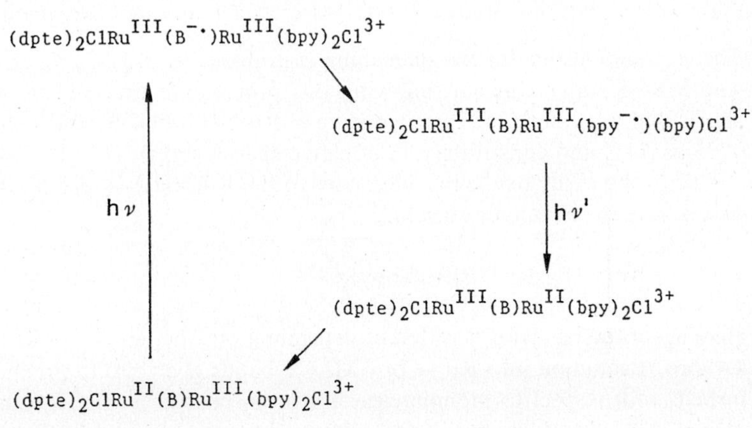

Scheme V

3.3.3 $(NH_3)_5Ru(B)Ru(bpy)_2Cl^{2+/3+}$ (B = bpa, bpe, 4,4′-bpy)

These binuclear complexes were studied by Curtis et al. [113]. The 2+ ions have the electronic structure

$$(NH_3)_5Ru^{II}(B)Ru^{II}(bpy)_2Cl^{2+}$$

The behavior of this type of complexes depends sharply on whether B = bpa or B = bpe, 4,4′-bpy. For the bpa complex, the emission properties match very closely those of the mononuclear $Ru(bpy)_2(bpa)Cl^+$ analogue, thus indicating that the Ru → bpy MLCT state is not quenched intramolecularly. For B = bpe, 4,4′-bpy, on the other hand, complete quenching of the emission is observed. This can be correlated with the presence of low-energy $(NH_3)_5Ru$ → B MBCT bands in the spectra of the bpe and 4,4′-bpy complexes (but not of the bpa one). This suggests the deactivation path shown in Scheme VI (where B = bpe, 4,4′-bpy).

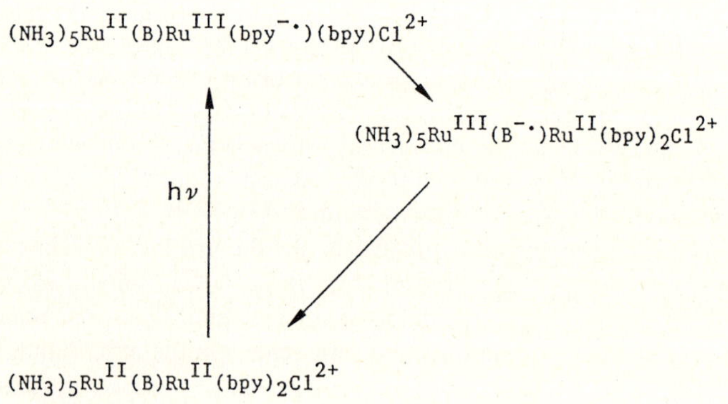

Scheme VI

The step responsible for the quenching is in this case an energy transfer process. The MBCT states reached following this process are expected to be extremely short-lived (e.g. the MLCT excited state of $Ru(NH_3)_5(4,4'\text{-}bpyH)^{3+}$ has a lifetime ≦30 ps [117] and constitute very effective energy sinks).

Due to the easily oxidizable character of the $Ru(NH_3)_5$— group, the 3+ complexes have the electronic structure

$$(NH_3)_5Ru^{III}(B)Ru^{II}(bpy)_2Cl^{3+}$$

Here again the behavior is different depending on whether B = bpa or B = bpe, 4,4′-bpy. Excitation into Ru → bpy MLCT bands gives rise to practically unperturbed (with respect to mononuclear $Ru(bpy)_2(bpa)Cl^+$) MLCT emission in the case of bpa, whereas no emission is observed for the bpe and 4,4′-bpy complexes. The intramolecular quenching observed in the case of bpe and 4,4′-bpy is ascribed

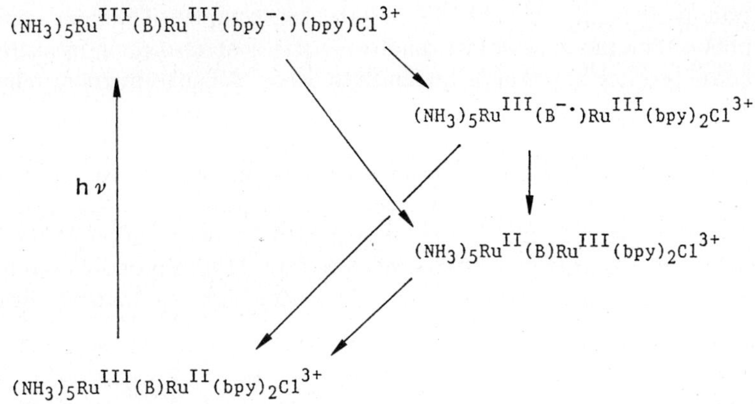

Scheme VII

by the authors [113] to two possible pathways, as shown in Scheme VII (where B = bpe, 4,4′-bpy). The first pathway is a ligand-to-ligand electron transfer populating a MBCT state, similar to that proposed for analogous Ru(II)—Ru(II) complexes (Scheme IV). The second is a ligand-to-remote-metal electron transfer process populating an IT state. No definite choice can be made between these two pathways on the basis of the experimental results. The fact that no quenching occurs in the bpa complex is also consistent with both hypotheses. In fact, the explanation may lie in the higher energy of the MBCT excited state of the bpa complex relative to the other ones, or, alternatively, in the smaller metal-metal electronic coupling provided by this bridging ligand [113].

3.3.4 $(bpy)_2(CO)Os(B)Os(phen)(dppe)Cl^{3+/4+}$ (B = bpa, 4,4′-bpy)

These systems were studied by Schanze, et al. [118, 119]. The 3+ complex has an electronic structure corresponding to

$$(bpy)_2(CO)Os^{II}(B)Os^{II}(phen)(dppe)Cl^{3+}$$

As in the cases discussed previously, the behavior is quite different depending on whether B = bpa or 4,4′-bpy.

With B = bpa, the visible absorption spectrum consists of partially overlapping Os → bpy and Os → phen MLCT bands, closely resembling those of the correspondent mononuclear analogues. Because of the different ancillary ligands in the two centers, the Os → phen bands are at lower energies than the Os → bpy ones. The same energy ordering is likely for the potentially emitting lowest MLCT states centered on the two fragments. Regardless of the excitation energy, predominant emission from the Os → phen MLCT state is always observed in this complex (as shown by comparison with the mononuclear analog $Os(phen)(dppe)Cl(bpa)^+$), while only a minor short-lived ($\tau \leqq 10$ ns at 300 K) component assignable to Os → bpy MLCT emission is present. Thus, the upper Os → bpy MLCT excited state is efficiently quenched in the binuclear complex. The same conclusion was obtained on the basis of laser photolysis transient absorption measurements. The behavior does not change qualitatively upon going from room temperature to 77 K.

As far as the mechanism of the quenching process is concerned, the obvious hypothesis of an energy transfer quenching mechanism was discarded by the authors on the basis of the excitation spectra showing that the quenching of the Os → bpy MLCT state is *not* accompanied by sensitization of the Os → phen MLCT emission. This led Schanze et al. [118, 119] to propose an intramolecular quenching pathway based on metal-to-metal electron transfer (Scheme VIII).

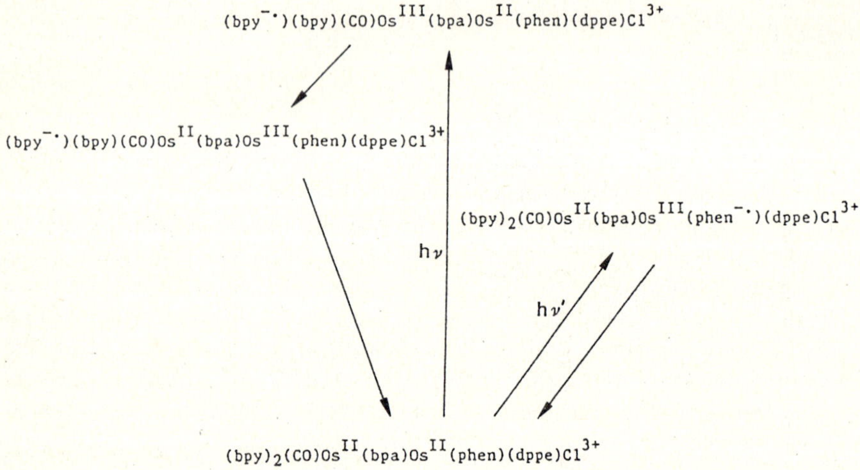

Scheme VIII

This process populates a remote MLCT that appears to relax to the ground state rather rapidly (estimated lifetime approx. 10^{-7} s) [119].

The finding that an electron transfer pathway is preferred over an energy transfer one in this system is of some relevance to the general question of the relative rates of these two types of processes. In this case, the states potentially involved in the two pathways (i.e. the remote Os → bpy and the "normal" Os → phen MLCT states) are at comparable energies and the Franck-Condon factors (reorganizational energies) for the two processes are presumably not too different. Thus, the observed behavior seems to point towards an inherently weaker electronic matrix element for energy relative to electron transfer in this system.

For the 4,4′-bpy-bridged complex, the behavior is strongly dependent on temperature [119]. The behavior in 77 K rigid matrices is essentially the same as for the bpa-bridged complex: quenching of the Os → bpy MLCT state, unquenched emission from the Os → phen MLCT state. Upon heating to temperatures above the glass transition of the matrix, however, the Os → phen MLCT emission is quenched too. This behavior is clarified by transient absorption measurements, that show that in fluid solutions an Os → (4,4′-bpy) MBCT state is formed. Thus, the proposed mechanism is as shown in Scheme IX. The process by which the Os → phen MLCT state converts to the lowest Os → (4,4′-bpy) MBCT state is a ligand-to-ligand electron transfer. That originating from the upper Os → bpy MLCT state is most probably a metal-to-metal electron transfer similar to that of Scheme VIII (not shown explicitly in Scheme IX). The fact that the low-energy deactivation path through the MBCT state does not operate at low temperature is explained, according to Schanze et al. [119] by the ability of the noncoplanar

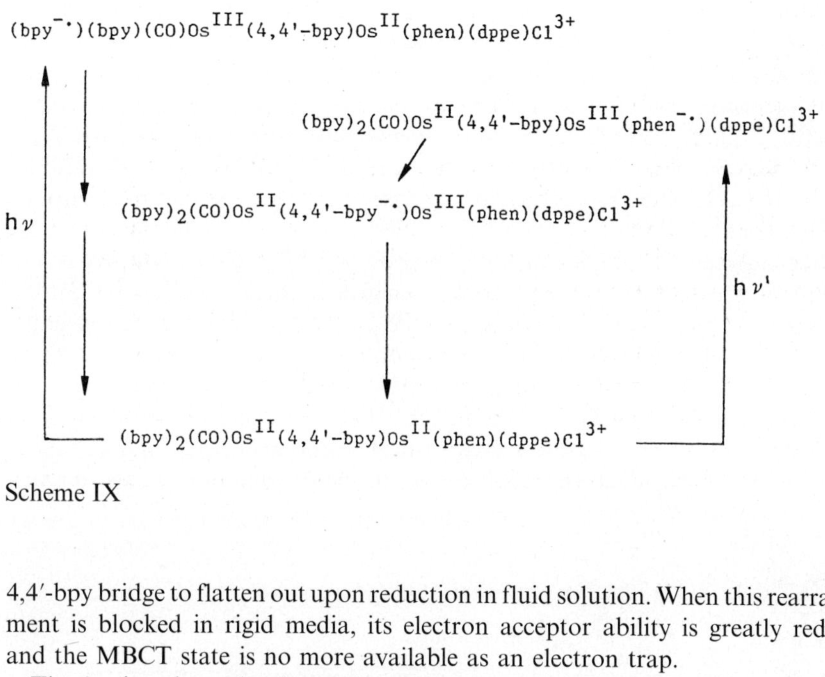

Scheme IX

4,4′-bpy bridge to flatten out upon reduction in fluid solution. When this rearrangement is blocked in rigid media, its electron acceptor ability is greatly reduced and the MBCT state is no more available as an electron trap.

The 4+ ions have the electronic structure

$$(bpy)_2(CO)Os^{II}(B)Os^{III}(phen)(dppe)Cl^{4+}$$

The visible spectrum is dominated by Os → bpy MLCT bands. In the near infrared, an Os(II) → Os(III) IT band is observed for B = 4,4′-bpy but not for B = bpa, presumably because of the poorer metal-metal electronic coupling provided by the latter bridge. The interesting photophysical result [119] is that, upon Os → bpy MLCT excitation, emission is observed with the same energy and lifetime as in the model $Os(bpy)_2(CO)(B)^{2+}$ mononuclear complexes. This indicates that the Os → bpy MLCT state is not quenched in the binuclear complex, despite the presence of the adjacent Os(III) site. It can be recalled that the presence of an oxidized metal was responsible for some of the proposed intramolecular quenching pathways in the $(dpte)_2ClRu^{II}(B)Ru^{III}(bpy)_2Cl^{3+}$ and $(NH_3)_5Ru^{III}(B)Ru^{II}(bpy)_2Cl^{4+}$

complexes discussed in Sects. 3.3.2 and 3.3.3, respectively. The lack for this system of an electron transfer quenching pathway proceeding via the low-energy IT state is discussed by Schanze et al. [119] in terms of the large exergonicity of the process (Marcus inverted region or energy-gap-law effects) and the large distance between the bpy$^{-\cdot}$ and Os(III) centers.

3.3.5 $(R_2\text{-bpy})(CO)_3Re(4,4'\text{-bpy})Re(R'_2\text{-bpy})(CO)_3^{2+}$ $(R, R' = NH_2, H, C_2H_5COO)$

This interesting series of complexes has been studied by Tapolsky et al. [120]. These complexes, in which both metal centers are in the +1 oxidation state, can be symmetric or asymmetric depending on the nature of the substituents on the 4,4′ positions at the bpy ligands. Relevant excited states in these systems are Re → (R_2-bpy) MLCT, Re → (R'_2-bpy) MLCT, and Re → (4,4′-bpy) MBCT excited states.

In the symmetric (R = R′) complexes the relative energy ordering of the MLCT and MBCT states depends on the substituents and being MLCT > MBCT for $R = R' = NH_2$, MLCT ≈ MBCT for R = R′ = H, and MBCT > MLCT for $R = R' = C_2H_5COO$, as determined by laser photolysis transient absorption measurements [120].

In the asymmetric complex in which R = H and $R' = C_2H_5COO$ (R'_2-bpy = = dec-bpy), the Re → (bpy) MLCT state is higher in energy than the Re → (dec-bpy) MLCT state. Excitation of the complex in the absorption region of either chromophore always gives rise to an emission and a transient absorption signal that can be assigned to Re → (dec-bpy) MLCT state. Two pathways could be responsible for the quenching of the upper MLCT state and the sensitization of the lowest one, namely, (i) a direct energy transfer process or (ii) the more complex "cascade" mechanism shown in Scheme X. In the cascade mechanism, the first

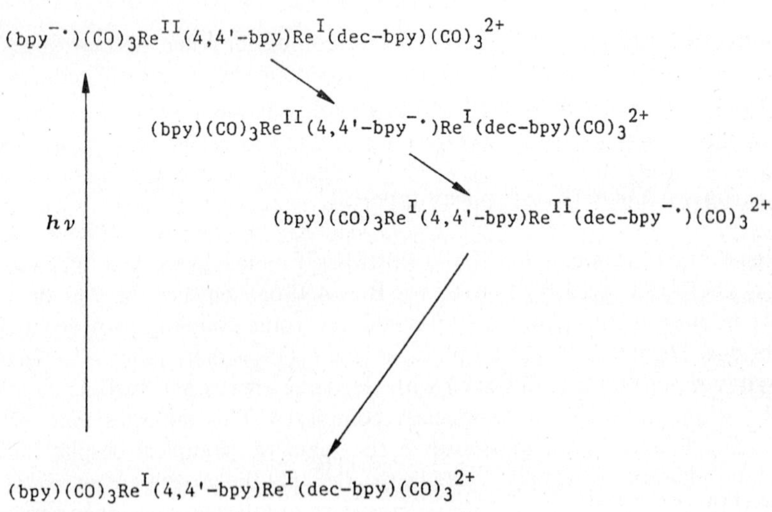

Scheme X

process is a ligand-to-ligand electron transfer generating a MBCT state, which then converts to the final MLCT state via energy transfer. The "cascade" mechanism is preferred by Tapolsky et al. [120] on the basis of the observation that intramolecular quenching and sensitization are completely blocked when 3,3′-$(CH_3)_2$-4,4′-bpy is substituted for the 4,4′-bpy bridging ligand. The new bridge is forced to be noncoplanar by the methyl substituents, being thus a poorer electron acceptor than 4,4′-bpy. Of course, the possibility that the mechanism is direct energy transfer and that the difference between the bridging ligands lies in their ability to provide electronic coupling cannot be definitely ruled out [120].

3.3.6 $X(NH_3)_4Ru—NC—Ru(bpy)_2—CN—Ru(NH_3)_4Y^{m+}$ and $X(NH_3)_4Ru—NC—Ru(bpy)_2(CN)^{n+}$ ($X = NH_3$, py; $Y = NH_3$, py; m = 4–6; n = 2, 3)

This series of binuclear and trinuclar complexes has been studied by Bignozzi et al. [121]. The possible combinations of X and Y ligands give rise to two binuclear and three trinuclear complex structures. Furthermore, the binuclear and trinuclear complexes can be isolated and studied in two and three different oxidation states, respectively, resulting in a total of thirteen complexes for this series. The various overall oxidation states reflect the individual oxidation states of the $Ru(NH_3)_4X$ and/or $Ru(NH_3)_4Y$ subunits in the complex. For example, the trinuclear complexes

$$py(NH_3)_4Ru—NC—Ru(bpy)_2—CN—Ru(NH_3)_5^{4+,5+,6+}$$

correspond to the following actual electronic structures:

$$py(NH_3)_4Ru^{II}—NC—Ru^{II}(bpy)_2—CN—Ru^{II}(NH_3)_5^{4+}$$
$$py(NH_3)_4Ru^{II}—NC—Ru^{II}(bpy)_2—CN—Ru^{III}(NH_3)_5^{5+}$$
$$py(NH_3)_4Ru^{III}—NC—Ru^{II}(bpy)_2—CN—Ru^{III}(NH_3)_5^{6+}$$

Contrary to what happened in the polynuclear complexes dealt with in the previous sections, in these complexes the bridging ligands are not chromophoric, as the excited states involving the cyanide ligands are at relatively high energy and can be safely neglected. Nonetheless, these complexes have a remarkable variety of charge transfer states in a photochemically interesting spectral range. Taking the $py(NH_3)_4Ru^{II}—NC—Ru^{II}(bpy)_2—CN—Ru^{III}(NH_3)_5^{5+}$ complex as an example, the following excited states are relevant:

1) Ru → bpy MLCT:
   $py(NH_3)_4Ru^{II}—NC—Ru^{III}(bpy^{-\cdot})(bpy)—CN—Ru^{III}(NH_3)_5^{5+}$
2) Ru → py MLCT:
   $(py^{-\cdot})(NH_3)_4Ru^{III}—NC—Ru^{II}(bpy)_2—CN—R^{III}(NH_3)_5^{5+}$
3) Ru → bpy remote MLCT:
   $py(NH_3)_4Ru^{III}—NC—Ru^{II}(bpy^{-\cdot})(bpy)—CN—Ru^{III}(NH_3)_5^{5+}$

4) Ru → Ru IT:
$py(NH_3)_4Ru^{II}—NC—Ru^{III}(bpy)_2—CN—Ru^{II}(NH_3)_5^{5+}$

5) Ru → Ru remote IT:
$py(NH_3)_4Ru^{III}—NC—Ru^{II}(bpy)_2—CN—Ru^{II}(NH_3)_5^{5+}$

The absortion spectra of these complexes are remarkably rich. In these spectra, transitions corresponding to the various types of excited states can be easily identified. For example, the resolution of the absorption spectrum of the above discussed complex into various types of transitions is shown in Fig. 5. By selective

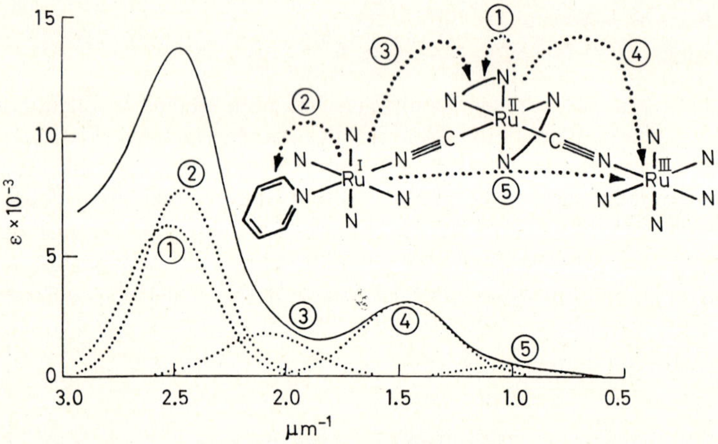

Fig. 5. Absorption spectrum of the trinuclear complex $py(NH_3)_4Ru—NC—Ru(bpy)_2—CN—Ru(NH_3)_5^{5+}$ and assignment of the component transitions to the various optical electron transfer processes

oxidation or reduction of the various sites in the molecule, the attribution of the various types of transitions is straightforward [121]. Of particular interest from the spectroscopic point of view is the direct observation of remote MLCT and of remote IT, indicating that, within the limits of an essentially localized description (Sects. 2.2 and 2.4), sizeable electronic coupling between the various sites is present in these systems. For the remote IT transition, the intensity appears to fit a super-exchange model for through-bond interaction between the terminal metal centers [122].

In all of the polynuclear complexes of this series, no emission can be detected following excitation in the Ru → bpy MLCT absorption band, indicating that efficient pathways are available for intramolecular quenching of the MLCT state of the $—Ru(bpy)_2—$ chromophore. These pathways can be easily identified on the basis of the states detected spectroscopically. For example, in the $[py(NH_3)_4Ru^{II}—NC—Ru^{II}(bpy)_2—CN—Ru^{III}(NH_3)_5]^{5+}$ complex, several electron-transfer quenching paths are available, as shown in Scheme XI. Analogous, though more simple, schemes can be elaborated for the other trinuclear complexes in the series. In so doing, it should be noticed that the remote IT state does not exist for fully oxidized (Ru(III)—Ru(II)—Ru(III)) and fully reduced (Ru(II)—Ru(II)—Ru(II)) complexes, and has the same energy (as a relaxed state) as the

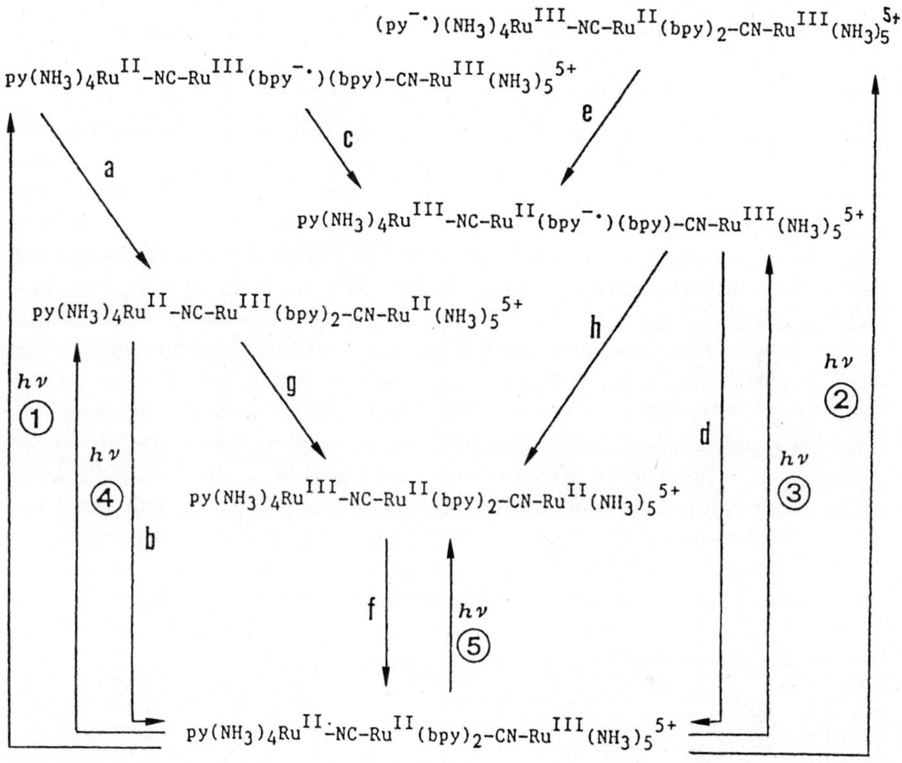

Scheme XI

ground state for symmetric semi-oxidized (Ru(II)—Ru(II)—Ru(III)) complexes. Thus, a sequence of processes of the type marked *a* and *b* in Scheme XI, i.e. proceeding via an IT state, holds for fully oxidized species, while the *c* and *d* sequence, proceeding via a remote MLCT state, constitutes the quenching pathway for fully reduced species. In Scheme XI, an intramolecular electron transfer pathway (initiated by process *e*) is also indicated for deactivation of the Ru → py MLCT excited state, although this type of state is expected to be intrinsically very short lived [117] and could not actually use this pathway. The extension of the above sketched arguments to the binuclear complexes is straightforward.

A noteworthy point is the similarity of the remote IT state of Scheme XI to the charge separated state of the donor-sensitizer-acceptor organic "triads" developed recently in a number of laboratories [40, 123, 124]. In the complexes of this series, no transient state is detected in nanosecond laser experiments, indicating that charge recombination steps (*b*, *d*, and *f* in Scheme XI) are extremely fast processes. The reasons for the lack of long-lived charge separation in these systems have been discussed in terms of thermodynamic and kinetic factors [121]. It should be pointed out that long-lived charge separation has recently been achieved by Danielson et al. [125] with a triad system containing a Ru-polypyridine sensitizer and organic donor and acceptor subunits.

3.3.7 $(NH_3)_5Ru(4\text{-}CNpy)Ru(bpy)_2(4\text{-}CNpy)Ru(NH_3)_5^{n+}$ (n = 6–8)

These complexes were synthesized and studied by Katz et al. [126] together with some analogs containings $Fe(CN_5^{2-,3-}$ instead of $Ru(NH_3)_5^{3+,2+}$. These complexes resemble structurally the trinuclear complexes of the previous section, having 4-cyanopyridine (nitrile-bound to the central Ru) instead of cyanide as the bridging groups. Here again, the three overall oxidation states are represented by Ru(II)—Ru(II)—Ru(II), Ru(II)—Ru(II)—Ru(III), and Ru(III)—Ru(II)—Ru(III) electronic structures. There is, however, a relevant difference with respect to the previous case, since the bridging ligands are chromophoric in these complexes. In particular, in the complexes that contain Ru(II) pentammine groups, low-energy $(NH_3)_5Ru \rightarrow$ (4-CNpy) MBCT states are present. The other relevant types of states in this series of complexes are, as in the previous case, Ru → bpy MLCT, Ru → bpy remote MLCT, Ru → Ru IT, and Ru → Ru remote IT states. All of these states show up in the absorption spectra of the appropriate complexes, except for the remote MLCT states whose absorption bands are probably hidden by more intense ones [126]. The IT and remote IT bands are less intense than in the cyanide case, indicating a smaller extent of metal-metal coupling.

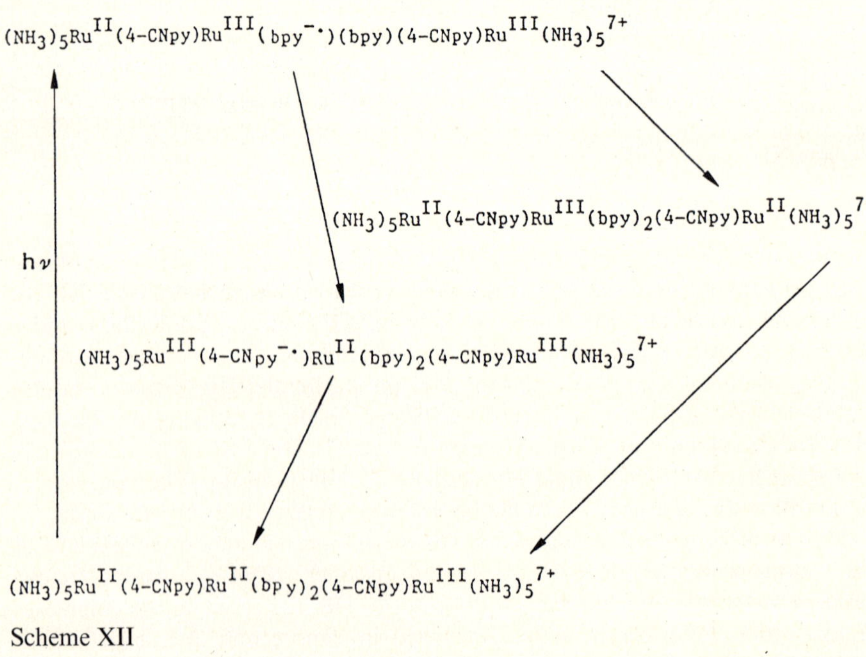

Scheme XII

The mononuclear analogue of the central chromophore, $Ru(bpy)_2(4\text{-}CNpy)_2^{2+}$, is nonemitting and quite photolabile at room temperature, but gives strong Ru → bpy MLCT emission at 77 K. This behavior is typical of thermal population of metal-centered d—d states from the lowest MLCT state. In the polynuclear complexes, regardless of the oxidation state, both the low-temperature emission and the room-temperature photoreactivity are strongly quenched, indicating the availability of efficient pathways for Ru → bpy MLCT excited state intramolecular deactivation. According to the authors, deactivation proceeds via the IT state for the fully oxidized species and via the MBCT state for the fully reduced species [126]. The two pathways are put together in Scheme XII, taking the $(NH_3)_5Ru(4\text{-}CNpy)Ru(bpy)_2(4\text{-}CNpy)Ru(NH_3)_5^{7+}$ semi-oxidized form as an example. As shown in Scheme XII, the step leading from the MLCT to the MBCT state is an energy transfer process involving a simultaneous two-electron change. The same result, however, could also be obtained by a sequence of two one-electron transfer processes (metal-to-metal and ligand-to-ligand) implying the intermediacy of a remote MLCT state (not shown in the scheme).

3.3.8 $(CN)(bpy)_2Ru—CN—Ru(bpy)_2—NC—Ru(bpy)_2(CN)^{m+}$ (m = 2, 3) and $(CN)(bpy)_2Ru—CN—Ru(bpy)_2(CN)^{n+}$ (n = 1, 2)

These complexes have been studied by Bignozzi et al. [127]. They contain two or three $—Ru(bpy)_2—$ units, with cyanides both as bridging and terminal ligands. The bonding mode of the bridging cyanides is determined by the synthetic procedure used. For the binuclear complex

$$(NC)Ru(bpy)_2—CN—Ru(bpy)_2(CN)^{n+}$$

there is obviously only one possible structure (although two possibilities would arise for analogous binuclear complexes with different polypyridine ligands on the two metal centers [127]) but for the trinuclear complexes three linkage isomers are possible, one of which has actually been synthesized and studied, i.e.

$$(NC)Ru(bpy)_2—CN—Ru(bpy)_2—NC—Ru(bpy)_2(CN)^{m+}$$

In the fully reduced species (n = 1 and m = 2), all the ruthenium centers are in the 2+ oxidation state. The various $—R^{II}(bpy)_2—$ chromophoric units, however, are not all identical, as they differ in the C- or N-bonded nature of the bridging cyanides. This affects to some extent the energies of the metal-to-ligand charge transfer (MLCT) states of the various $Ru(bpy)_2^{2+}$ chromophores, that decrease (by approx. $0.15\ \mu m^{-1}$ for each step) in the order NC—Ru—CN > > NC—Ru—NC > CN—Ru—NC. In these Ru(II)—Ru(II)—Ru(II) systems, a single emission attributable to the lowest energy chromophore is observed. Although no selective population of the various MLCT excited states is possible in these systems due to overlapping absorption bands, the exact correspondence between excitation and absorption spectra in these complexes points towards a very efficient intramolecular energy transfer process from the higher-energy chromophores to the lowest emitting one [127]. The photophysical behavior of these polychromophoric Ru(II) species is represented in Scheme XIII, using the

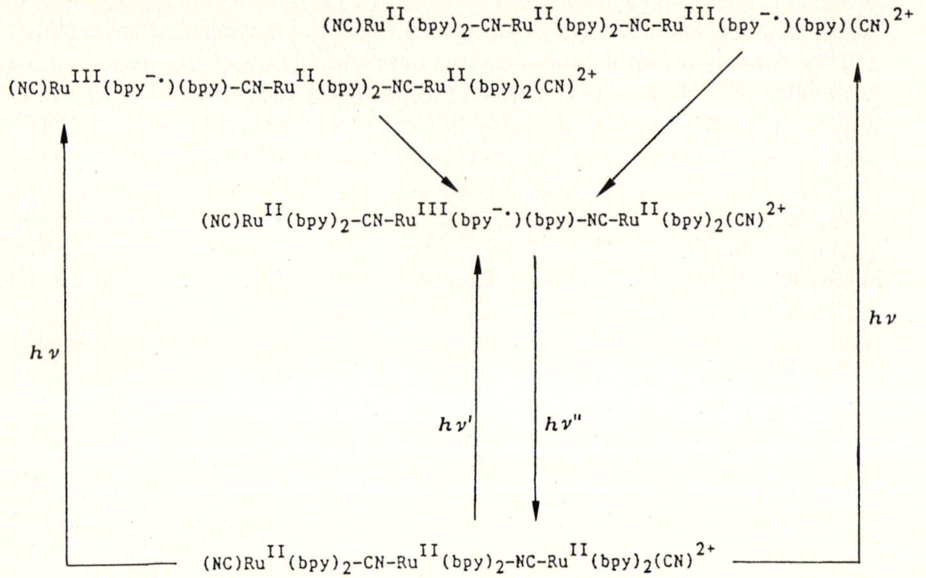

Scheme XIII

trinuclear complex as an example. The steps converting the higher MLCT states to the lowest one are most likely energy transfer processes, although a two-step metal-to-metal, ligand-to-ligand electron transfer sequence cannot be strictly ruled out. As far as the mechanism of energy transfer is concerned, a singlet-singlet process is unlikely in view of the fast and efficient intersystem crossing that characterizes Ru(II) polypyridine complexes [128]. For the more plausible triplet-triplet pathway, on the other hand, both dipole-dipole (Forster-type) and exchange (Dexter-type) mechanisms (Sect. 2.3) can be considered. A Forster mechanism has been shown [71, 73] to account for the energy transfer processes observed in binuclear polypyridine compounds in which the polypyridine ligands on the two metal centers are connected by polymethylene chains (Sects. 3.1.2 and 3.1.3). Realistic calculations using the Forster relationships are not possible in this case because of the unavailability of relevant parameters. Overall, given the strong electronic coupling provided by the bridging cyanide in these (see below) and similar complexes (Sects. 3.3.6 and 3.3.7), an exchange Dexter-type mechanism seems to be more likely in these systems [127].

The related trinuclear complex

$$(NC)Ru^{II}(bpy)_2{-}CN{-}Ru^{II}(dc\text{-}bpy)_2{-}NC{-}Ru^{II}(bpy)_2(CN)^{2-}$$

has been recently synthesized and studied [129]. This complex is better suited to investigate intramolecular energy transfer since the presence of the carboxylic groups on the bipyridines of the central chromophoric unit have the effect of further lowering the MLCT energy of this unit. This leads to sizeable shifts in MLCT absorption and thus to the possibility to address the individual chromophores with light of different wavelength. Also in this case, emission from the central chromophore is observed with constant efficiency, independent on the nature of excited chromophore.

The results show that in these systems fast (subnanosecond) intramolecular energy transfer between adjacent $—Ru^{II}(bpy)_2—$ chromophores leads to 100% efficient population of the lowest energy chromophore. It has been remarked [127, 129] that polychromophoric systems of this type exhibit an "antenna effect" similar (though probably different in mechanism) to that by which the antenna pigments greatly increase the effective absorption cross-section of the special pair in natural photosynthetic reaction centers [59].

One-electron oxidation of the above discussed binuclear and trinuclear complexes gives mixed-valence species of the type

$$(NC)Ru^{II}(bpy)_2—CN—Ru^{III}(bpy)_2(CN)^{2+}$$

and

$$(NC)Ru^{II}(bpy)_2—CN—Ru^{III}(bpy)_2—NC—Ru^{II}(bpy)_2(CN)^{3+}$$

in which the metal center containing N-bonded bridging cyanide(s) is oxidized [127]. Besides the MLCT bands characteristic of the $—Ru^{II}(bpy)_2—$ units, these com-

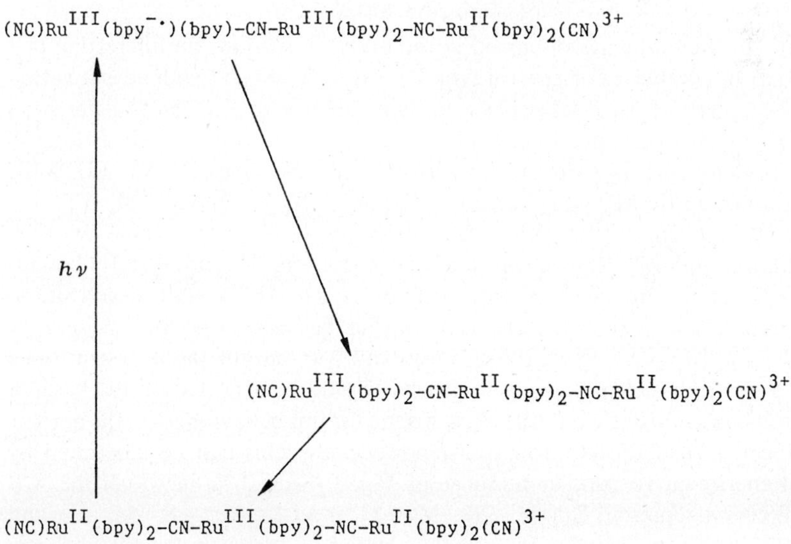

Scheme XIV

plexes exhibit Ru(II) → Ru(III) IT bands in the near infrared. The intensity of these bands points toward a relatively large degree of electronic coupling between the metals (borderline between Class II and III [63], Sect. 2.4). The binuclear and trinuclear mixed valence complexes do not emit upon MLCT excitation. This indicates that intramolecular electron transfer quenching of the MLCT states takes place, as shown in Scheme XIV for the trinuclear complex. It should be recalled that an analogous intramolecular electron transfer pathway was not followed in the Os(II)—Os(III) complex discussed in Sect. 3.3.4. The inefficiency of this process in that case was explained by Schanze et al. [119] on the basis of the large exergonicity and/or the large distance (poor communication) between the redox sites involved. The comparison between the two cases indicates that the second factor is more plausible. In fact, the main difference between the two systems lies in the shorter and more delocalizing cyanide bridge relative to the longer and more insulating 4,4′-bpy and bpa bridges.

3.3.9 Remarks

There are two conceptually different ways of looking at the deactivation of an electronically excited polynuclear complex. In a "molecular" view, the deactivation is seen in terms of radiationless (and radiative) transitions from upper to lower excited states of the polynuclear complex, much in the same way as one would do with a simple molecule. In a "supramolecular" view (Sect. 2.4), one can describe deactivation in terms of intramolecular energy or electron transfer processes between the various components of the polynuclear complex. In the discussion of the experimental studies carried out in Sects. 3.3.1–3.3.8, the supramolecular approach has been largely followed. From this point of view, the systems discussed have been found to exhibit a remarkable variety of energy and electron transfer deactivation pathways (Schemes III–XIV).

In many of the pathways discussed in the previous sections an interesting fact is observed: in complexes of general type $L_a—M_a—B—M_b—L_b$, the deactivation of a $M_a \rightarrow L_a$ MLCT state takes place via two subsequent electron transfer steps involving another component (the bridging ligand or the other metal). Examples of such sequences are $L_a \rightarrow B$ followed by $B \rightarrow M_a$ (Schemes IV, VI, VII, XII), $L_a \rightarrow M_b$ followed by $M_b \rightarrow M_a$ (Schemes VII, XI, XII, XIV), $M_b \rightarrow M_a$ followed by $L_a \rightarrow M_b$ (Schemes VIII, XI). The intriguing point is that these two-step sequences, that involve electron transfer between remote (i.e., not directly bound) sites, are faster by orders of magnitude than direct MLCT state deactivation, that corresponds to a one-step electron transfer between adjacent (i.e. directly bound) L_a and M_a sites. Thus, the shortest way is often not the fastest in these systems. Since this is clearly opposite of what would be expected on the basis of the electronic factors (Sect. 2.1), the likely reason for this behavior lies in the nuclear factors (Sect. 2.1) of the electron transfer processes. This can be illustrated by potential energy curves similar to those of Fig. 3 (Sect. 2.2), in which the A.B and A*.B curves represent $L_a—M_a—B—M_b—L_b$ and $L_a^-—M_a^+—B—M_b—L_b$ and the $A^+.B^-$ curve represents an intermediate state in a two-step deactivation, e.g. $L_a—M_a^+—B—M_b^-—L_b$. In this picture, the direct deactivation corresponds to electron transfer in the Marcus inverted region (Sect. 2.1), whereas each step of the two-step sequence, due to the smaller exergonicity and larger reorganizational energy, is an almost activationless electron transfer process.

Of course, complex energy and electron transfer pathways and mechanistic problems similar to those encountered in Sect. 3.3 are common to many supramolecular systems other than polynuclear complexes. Extensive work has been carried out on organic donor-acceptor systems (dyads, triads, etc.) [40, 50, 56, 62, 130, 131]. As a consequence of these studies, a great deal of basic knowledge on energy and alectron transfer kinetics has been gained and important progress towards efficient photoinduced charge separation has been made. There are, on the other hand, several very interesting studies on systems that are intermediate between the organic donor-acceptor molecules and the polynuclear complexes discussed here, in that they contain an inorganic light absorbing chromophore (e.g. a Ru(II) polypyridine complex) and covalently bound organic electron acceptors (e.g., quinones, bipyridinium ions) or donors (e.g. phenothiazine) [132–138]. Also, some systems that contain an organic light absorbing chromophore bound to inorganic electron acceptors or donors have been recently investigated [139, 140]. Although the basic problems involved in these chromophore-quencher complexes are quite similar to those found in polynuclear complexes, these systems are not discussed in this review because of space limitations.

3.4 Photochemical Pathways in Polynuclear Complexes

Under this heading are discussed a number of studies in which the main emphasis is placed on *photochemical reactions* of polynuclear complexes. Interesting aspects of these studies are (i) the quantitative changes in the photoreactivity of the fragments induced by inclusion into the polynuclear structure, and (ii) the possibility to induce new types of photochemical reactions, not exhibited by the separate fragments. Point (i) is mainly related to quenching or sensitization of the photoreactions of the fragments by means of intramolecular energy transfer:

$$L_aM_a{-}B{-}M_bL_b \xrightarrow{h\nu} L_aM_a^*{-}B{-}M_bL_b \downarrow \text{products}$$

$$L_aM_a^*{-}B{-}M_bL_b \rightarrow L_aM_a{-}B{-}M_b^*L_b \downarrow \text{products}$$

Point (ii) usually derives from the kinetic lability of the reduced form of some metal-containing fragments that can be produced following localized (e.g. MBCT) excitation and intramolecular electron transfer

$$L_aM_a{-}B{-}M_bL_b \xrightarrow{h\nu} L_aM_a^+{-}(B^{-\cdot}){-}M_bL_b$$

$$L_aM_a^+-(B^{-\cdot})-M_bL_b \rightarrow L_aM_a^+-B-M_b^-L_b \atop \downarrow \atop \text{products}$$

or by direct intervalence transfer (IT) excitation

$$L_aM_a-B-M_bL_b \xrightarrow{hv} L_aM_a^+-B-M_b^-L_b \atop \downarrow \atop \text{products}$$

Intervalence transfer excitation involving two metals of a binuclear complex has obvious similarities with excitation of "outer-sphere charge transfer" transitions in systems in which two metal complexes are held together by non-covalent (e.g. electrostatic) interactions. The photochemistry of ion-paired transition metal complexes has been the subject of considerable interest [65, 141, 142] and is reviewed by Vogler in another chapter of this volume.

3.4.1 Intramolecular Quenching and Sensitization. *$(CN)_5Co—NC—Co(NH_3)_5$ and $(CN)_5Co—CN—Co(NH_3)_5$*

An interesting study on this system was performed by Nishizawa and Ford [143]. The two complexes are linkage isomers that differ in the bonding mode of the bridging cyanide. In spite of their similarity, the two isomers exhibit quite different photochemical behavior. For both $(CN)_5Co—NC—Co(NH_3)_5$ and $(CN)_5Co—CN—Co(NH_3)_5$, the absorption spectrum is a simple superimposition of the spectra of suitable model subunits (e.g. $Co(CN)_5(CH_3CN)^{2-}$ and $Co(NH_3)_5CN^{2+}$, and $Co(CN)_6^{3-}$ and $Co(NH_3)_5(CH_3CN)^{3+}$, respectively), pointing towards relatively small mutual perturbation of the fragments in the binuclear complexes. The spectra are such that, in both cases, selective excitation of the $Co(CN)_5$-based chromophore can be achieved with ultraviolet light and of the $Co(NH_3)_5$-based one with visible light.

For $(CN)_5Co—NC—Co(NH_3)_5$, excitation in the ultraviolet range gives rise to a very efficient photocleavage reaction (Eq. 25), whereas visible excitation

$$(CN)_5Co\text{-}NC\text{-}Co(NH_3)_5 + H_2O \xrightarrow[\Phi=0.28]{hv(366\,nm)}$$

$$Co(CN)_5H_2O^{2-} + Co(NH_3)_5CN^{2+} \qquad (25)$$

has very little photochemical consequence ($\Phi \approx 10^{-3}$) [143]. This behavior is what would be expected on the basis of the known photoreactivity of the fragments (high photosubstitution yields in LF photochemistry of pentacyano-cobaltate(III) complexes, very low photoreaction yields in LF photochemistry of pentammine-cobalt(III) complexes [144] and points towards an independent behavior of the two fragments in the binuclear complex.

For the linkage isomer $(CN)_5Co—CN—Co(NH_3)_5$, on the contrary, both ultraviolet and visible excitation only give rise to small photoreaction quantum yields (approx. 10^{-3} and 10^{-5}, respectively). Thus, the expected photoreactivity of the $Co(CN)_5$-based chromophore appears to be efficiently quenched in this binuclear complex. The process responsible for this quenching is, as suggested by Nishizawa and Ford [143], an intramolecular energy transfer converting the lowest LF triplet of the $Co(CN)_5$—CN— fragment to the lowest LF triplet of the —CN—$Co(NH_3)_5$ fragment.

The difference in behavior between the two linkage isomers could in principle be attributed to differences in either the rate of bond cleavage in the excited ($Co(CN)_5(CN)$— ((CN) = —CN— or —NC—) chromophore (Eq. 27) or the rate of intramolecular energy transfer (Eq. 28). The authors favor the latter

$$(CN)_5Co(CN)Co(NH_3)_5 \xrightarrow{h\nu} (CN)_5Co^*(CN)Co(NH_3)_5 \quad (26)$$

$$(CN)_5Co^*(CN)Co(NH_3)_5 \xrightarrow{H_2O} Co(CN)_5H_2O^{2-} + Co(NH_3)_5(CN)^{2+} \quad (27)$$

$$(CN)_5Co^*(CN)Co(NH_3)_5 \rightarrow (CN)_5Co(CN)Co^*(NH_3)_5 \quad (28)$$

explanation and tentatively attribute the difference in energy transfer rate constants to symmetry factors arising from the different tetragonal splittings of the *d*-orbitals in the two isomers [143]. Whatever the actual mechanism may be, the fact that a simple inversion in the coordination mode of the bridging cyanide brings about a large difference in energy transfer rates is remarkable.

$(CN)_5Fe(B)Co(CN)_5^{5-}$, $(CN)_5Fe(B)Rh(NH_3)_5$, and $(NH_3)_5Ru(B)Rh(NH_3)_5^{5+}$

These complexes have been studied by Gelroth et al. [145] and Moore et al. [146]. The author's aim was to have in the same molecule a highly absorbing and photochemically stable chromophore (($(CN)_5Fe(B)$- or $(NH_3)_5Ru(B)$-, with characteristic Fe → B or Ru → B MBCT bands) and a nonabsorbing and photolabile fragment (—$(B)Co(CN)_5$ or —$(B)Rh(NH_3)_5$), and to look for intramolecular energy transfer from the light absorbing chromophore to the reactive fragment. These systems, however, display a much more complex photochemical behavior than expected on these simple grounds.

All the complexes containing the $(CN)_5Fe(B)$— chromophore undergo with high quantum yield (Φ = 0.6–0.01) a bleaching of the MBCT band [146], indicating the occurrence of a photocleavage reaction at the Fe—B bond (e.g. Eq. 29)

$$(CN)_5Fe(B)Co(CN)_5^{5-} \xrightarrow[H_2O]{h\nu} Fe(CN)_5H_2O^{3-} + Co(CN)_5B^{2-} \quad (29)$$

This result is unexpected, since mononuclear $(CN)_5Fe(B)^{3-}$ species of comparable MLCT energy are known to be virtually photostable [147]. Apparently, (i) the Fe → B MBCT state has a different reactivity in the binuclear complexes and (ii) the MBCT state is not quenched by intramolecular energy transfer. A number of tentative explanations for these results are discussed by Moore et al. [146].

For the $(NH_3)_5Ru(B)Rh(NH_3)_5^{5+}$ complexes, the photochemical behavior was found to be dependent on the nature of the bridging ligand B [145, 146]. With B = pz, practically no photochemistry is observed ($\Phi \approx 10^{-5}$). With B = 4-CNpy, moderate yields ($\Phi \approx 10^{-2}$–10^{-4}) of Rh—B bond cleavage (Eq. 30) were observed

$$(NH_3)_5Ru(B)Rh(NH_3)_5^{5+} \xrightarrow[H_2O]{h\nu} Ru(NH_3)_5B^{2+} + Rh(NH_3)_5H_2O^{3+} \tag{30}$$

With B = 4,4′-bpy competitive Rh—B (Eq. 30) and Ru—B (Eq. 31) bond cleavage of moderate quantum yields ($\Phi \approx 10^{-2}$–10^{-3}) was observed. Since the Rh–B

$$(NH_3)_5Ru(B)Rh(NH_3)_5^{5+} \xrightarrow[H_2O]{h\nu} Ru(NH_3)_5H_2O^{2+} + Rh(NH_3)_5B^{3+} \tag{31}$$

bond cleavage can be considered a reaction characteristic of LF states of Rh(III) [148], its observation upon Ru → BMBCT excitation is taken as an indication of intramolecular energy transfer from the Ru-based to the Rh-based fragments [146]. As discussed by the authors, however, there are problems in accounting for the dependence of the behavior on B, as the energy of the MBCT donor state follows the order 4-CNpy > pz > 4,4′-bpy whereas that of the energy transfer efficiency would be 4-CNpy ≈ 4,4′-bpy ≫ pz. An alternative mechanism based on B → Rh electron transfer followed by reaction at the labile Rh(II) center (a mechanism similar to those discussed in Sects. 3.4.2 and 3.4.3) is also discussed, but considered less plausible, by Moore, et al. [146].

The strategy of putting together in a polynuclear complex a stable chromophore and a photoreactive fragment has also been followed by McQeen and Petersen [149] in their recent study of $(bpy)_2Ru(B)Rh(PPh_3)_2H_2^{3+}$ (B = bpm, dpp, dpq). This complex gives rise to a typical MBCT emission following excitation in a wide spectral range including the low-energy Ru → B MBCT band. On the other hand, the complex undergoes reductive elimination of molecular hydrogen (a reaction typical of the Rh-centered fragment), but only upon excitation at shorter wavelengths where Rh → B MBCT bands are present. Thus, in this system there seems to be no need to invoke intramolecular energy transfer to explain qualitatively the observed behavior.

3.4.2 Photoinduced Electron Transfer Photochemistry. *$(CN)_5Fe(B)Co(NH_3)_5$ (B = pz, Mepz)*

These interesting systems were studied several years ago by Malin et al. [150]. The complexes were formed in situ in a stopped-flow apparatus by reaction of $Fe(CN)_5H_2O^{3-}$ and $Co(NH_3)_5B^{3+}$. The formation reaction gives the thermodynamically unstable Fe(II)—Co(III) species, that further reacts to give $Fe(CN)_5B^{2-}$ and free Co(II). This decomposition reaction actually occurs via consecutive intramolecular electron transfer (e.g. Eq. 32) and decomposition

$$(CN)_5Fe^{II}(B)Co^{III}(NH_3)_5 \rightarrow (CN)_5Fe^{III}(B)Co^{II}(NH_3)_5 \tag{32}$$

of the labile Co(II) center formed (Eq. 33). The electron transfer process is

$$(CN)_5Fe^{III}(B)Co^{II}(NH_3)_5 \rightarrow Fe(CN)_5B^{2-} + Co^{2+}_{aq} + 5\ NH_3 \qquad (33)$$

rate determining, with measured rates of the order of 10^{-1}–10^{-2} s^{-1}.

Interestingly, Malin et al. [150] observed that the intramolecular electron transfer reaction was accelerated by light. The quantum yields for the photoinduced reaction were practically unitary. Since the excitation with visible light corresponds to Fe → B MBCT transitions, the mechanism of the photochemical reaction involves a B → Co intramolecular electron transfer following excitation (Scheme XV). Looking at the results of Malin et al. [150] from the standpoint of electron transfer theory (Sect. 2.1), the large acceleration factor of the excited-state process (estimated $k > 10^9$ s^{-1}) over the thermal one can be traced back to both electronic and nuclear factors. In fact, (i) the electronic coupling is likely to be stronger between B and Co(III) than between Fe(II) and Co(III) and (ii) the larger exergonicity of the excited-state reaction can better overcome the large reorganizational energies [151] generally associated to Co(III)/Co(II) reduction. The fact, demonstrated by the unitary quantum yields, that back electron transfer from B to Fe(III) (MBCT excited state deactivation) is much slower than forward electron transfer from B to Co(III) is also noteworthy. The simplest explanation could be based

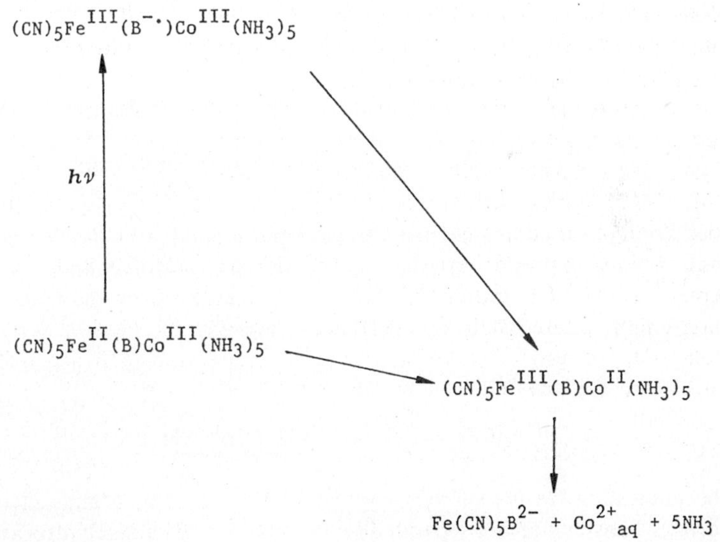

Scheme XV

on the different reorganizational energy of the two processes. Given the large exergonicity of both processes, the process involving the Fe(II)/Fe(III) couple could be in the inverted Marcus region, while that involving the Co(II)/Co(III) couple (with large reorganizational energy) could be in the almost activationless region. Of course, the picture would become more complicated if excited states (e.g. low-spin) of the Co(II) were involved in the electron transfer process (see Sect. 3.4.3).

$$Cu[\pi\text{-}(CH_2{=}CH{-}R{-}NH_2)]Co(NH_3)_5^{4+} \text{ and}$$
$$Cu[\pi\text{-}(CH_2{=}CH{-}R{-}CH_2py)]Co(NH_3)_5^{4+}$$
$$(R{=}(CH_2)_n,\ n = 1\text{–}4, 6, 8);$$
$$Cu[\pi\text{-}(C(CH_3)_2{=}CH{-}R'{-}NH_2)]Co(NH_3)_5^{4+}$$
$$Cu[\pi\text{-}(C(CH_3)_2{=}CH{-}R'{-}CH_2py)]Co(NH_3)_5^{4+}$$
$$(R' = [(CH_2)_2C(CH_3){=}CH))_nCH_2,\ n = \mathbf{1}, \mathbf{2})$$

This remarkable series of binuclear complexes has been studied by Norton and Hurst [152], as an extension of previous work on similar systems [153]. In these complexes, Cu(I) is bound in a π-fashion to the olefinic group on one end of the bridge, while the $Co(NH_3)_5^{3+}$ fragment is bound to an amino or a pyridine group at the other end. Saturated polymethylene (R) or partially unsaturated polyisoprene (R′) chains of various lengths provide the chemical link between the two ends of the bridge. These complexes are formed in situ by reaction of cuprous ions with the pentamminecobalt(III) complex of the bridging ligand. In these systems, thermal electron transfer from Cu(I) to Co(III), though thermodynamically allowed, is completely negligible for kinetic reasons.

The binuclear complexes display an intense band in the near ultraviolet region that is assigned to Cu → olefin MBCT transitions (e.g. Eq. 34).

$$Cu^{I}[\pi{-}(CH_2{=}CH{-}R{-}NH_2)]Co^{III}(NH_3)_5^{4+} \xrightarrow{h\nu}$$
$$Cu^{II}[\pi{-}(^{-\cdot}CH_2{=}CH{-}R{-}NH_2)]Co^{III}(NH_3)_5^{4+} \qquad (34)$$

Irradiation of the complexes into this absorption band gives rise to a redox decomposition reaction producing aqueous Co^{2+} [152]. The primary and rate determining step is thought to be intramolecular electron transfer from the reduced ethylenic group of the excited state to Co(III) (e.g. Eq. 35)

$$Cu^{II}[\pi\text{-}(^{-\cdot}CH_2{=}CH{-}R{-}NH_2)]Co^{III}(NH_3)_5^{4+} \rightarrow$$
$$\rightarrow Cu^{II}[\pi\text{-}(CH_2{=}CH{-}R{-}NH_2)]Co^{II}(NH_3)_5^{4+} \qquad (35)$$

The interesting feature of these systems is that the two groups involved in the electron transfer step are separated by more or less extended hydrocarbon chains, so that the effect of the chain length and nature on the electron transfer rates can be investigated. The main observations made by Norton and Hurst are [152]: (i) for the $Cu[\pi\text{-}(CH_2{=}CH{-}R{-}NH_2)]Co(NH_3)_5^{4+}$ and $Cu[\pi\text{-}(CH_2{=}CH{-}R{-}CH_2py)]Co(NH_3)_5^{4+}$ ($R = (CH_2)_n$, n = 1–4, 6, 8) complexes, the quantum yields decrease regularly with increasing number of methylene units in the R chain, being relatively high (0.65 and 0.15) for n = 1 and becoming immeasurably small for $n \geq 6$; (ii) for the $Cu[\pi\text{-}(C(CH_3)_2 = CH{-}R'{-}NH_2)]Co(NH_3)_5^{4+}$ and

$Cu[\pi\text{-}((CH_3)_2{=}CH{-}R'{-}CH_2py)]Co(NH_3)_5^{4+}$ ($R'{=}[(CH_2)_2C(CH_3){=}CH]_nCH_2$, $n = 1, 2$) complexes, the quantum yields are close to unity and independent on the number of isoprenic groups in the R′ chain; (iii) when the quantum yields are sufficiently high to permit laser photolysis measurements, the rate constants of the intramolecular electron transfer step are greater than 10^8 s^{-1}.

In the interpretation of result (i), the authors assume that these flexible molecules adopt a fully extended equilibrium conformation in the experimental conditions used (this point is supported by molecular mechanics calculations, and by comparisons of quantum yields with those of a related rigid systems of known center-to-center distance) [152]. Thus, the increase in the number of methylene groups translates into an increase in center-to-center distance. The results of Norton and Hurst represent one of the earliest clear experimental examples of distance dependence of electron transfer kinetics in covalently linked donor-acceptor systems. Following extensive work on organic donor-acceptor systems [40, 50, 56, 62, 131], it is now widely recognized that electron transfer rates are expected to decrease exponentially with distance regardless of whether through-space interactions are only considered or a through-bond (Sect. 2.1) mechanism is assumed. For a recent, partially inorganic study taking a similar basic approach, the reader is referred to the work of Schanze and Sauer [134] on photoinduced electron transfer in $Ru(bpy)_3^{2+}$-quinone donor-acceptor pairs linked by polyproline chains of various lengths.

The strong accelerating effect of the insertion of double bonds in the methylene chain (point (ii)) is interpreted by the authors in terms of delocalized interactions between the remote redox centers proceeding via $\pi{-}\pi$ interactions along the chain [152]. The fact that excited-state deactivation competes with the fast (point (iii)) intramolecular electron transfer processes indicates that the original Cu → olefin MBCT state is a rather short-lived one. This led Norton and Hurst to suggest a singlet state as the reactive state in these systems [152].

3.4.3 Intervalence Transfer Photochemistry. *$(CN)_5M_a{-}CN{-}M_b(NH_3)_5^-$ (M_a = Fe, Ru, Os; M_b = Cr, Co, Os)*

These binuclear complexes have been studied in some detail by Vogler et al. [141, 154, 155]. In these complexes, M_a is in the 2+ and M_b in the 3+ oxidation state. All the complexes exhibit, in addition to absorption features characteristic of the component subunits, an intense $M_a \rightarrow M_b$ intervalence transfer (IT) transition (Eq. 36) in the visible. The results obtained by

$$(CN)_5M_a^{II}{-}CN{-}M_b^{III}(NH_3)_5^- \xrightarrow{h\nu} (CN)_5M_a^{III}{-}CN{-}M_b^{II}(NH_3)_5^- \quad (36)$$

irradiating the complexes in the wavelength range of this band differ sharply depending on the nature of M_b.

When M_b = Os [155], the $-CN-Os^{II}(NH_3)_5$ unit present in the IT excited state is relatively inert, so that excitation is followed by efficient back electron transfer reaction (e.g. Eqs. 37, 38). In this respect, these complexes

$$(CN)_5Ru^{II}-CN-Os^{III}(NH_3)_5^- \xrightarrow{h\nu} (CN)_5Ru^{III}-CN-Os^{II}(NH_3)_5^- \quad (37)$$

$$(CN)_5Ru^{III}-CN-Os^{II}(NH_3)_5^- \rightarrow (CN)_5Ru^{II}-CN-Os^{III}(NH_3)_5^- \quad (38)$$

behave very similarly to the systems that deactivate via intramolecular electron transfer pathways described in Sect. 3.3.

When M_b = Cr [155] or Co [154], the $-CN-M_b^{II}(NH_3)_5$ unit present in the IT excited state is very labile, so that an efficient redox photo decomposition reaction is observed (e.g. Eqs. 39, 40). The reported photosensitivity of the related

$$(CN)_5Ru^{II}-CN-Co^{III}(NH_3)_5^- \xrightarrow{h\nu} (CN)_5Ru^{III}-CN-Co^{II}(NH_3)_5^- \quad (39)$$

$$(CN)_5Ru^{III}-CN-Co^{II}(NH_3)_5^- \rightarrow Ru(CN)_6^{3-} + Co^{2+} + 5\,NH_3 \quad (40)$$

$(CN)_5Ru-CN-Co(NH_3)(en)_2^-$ complex [156] most probably reflects the same type of reactivity. In this respect, these complexes resemble the systems described in Sect. 3.4.2. The main difference is, of course, that here the metal-to-metal electron transfer state is reached directly by an optical process (IT excitation), whereas in the above described cases this was accomplished by a sequence of MBCT excitation and intramolecular thermal electron transfer.

$(CN)_5M-CN-Co(CN)_5^{6-}$ (M = Fe, Ru, Os)

These complexes were also studied by Vogler et al. [155, 157]. As some of the complexes discussed in the previous section, these complexes contain M(II) and Co(III), and intense M → Co IT transitions are observed in the spectra. The behavior obtained upon IT excitation is, however, different. This arises from the different properties of the Co(II) ammine and cyano fragments present in the IT state. Contrary to the $-CN-Co^{II}(NH_3)_5$ fragment (which, as a high-spin complex, tends to undergo complete dissociation into aquo metal and ligands), the $-CN-Co^{II}(CN)_5$ unit is a strongly Jahn-Teller-distorted low-spin species, that easily dissociates one ligand retaining a pentacoordinated structure. The consequence is that in these cases IT excitation is followed by dissociation of the binuclear complex into mononuclear redox products (e.g. Eqs. 41, 42). In deaerated solutions, complete regeneration of the binuclear

$$(CN)_5Ru^{II}-CN-Co^{III}(CN)_5^{6-} \xrightarrow{h\nu} (CN)_5Ru^{III}-CN-Co^{II}(CN)_5^{6-} \quad (41)$$

$$(CN)_5Ru^{III}-CN-Co^{II}(CN)_5^{6-} \rightarrow Ru(CN)_6^{3-} + Co(CN)_5^{3-} \quad (42)$$

complex occurs by a thermal inner-sphere redox reaction (Eq. 43). In the

$$Ru(CN)_6^{3-} + Co(CN)_5^{3-} \rightarrow (CN)_5Ru^{II}—CN—Co^{III}(CN)_5^{6-} \quad (43)$$

presence of air, on the contrary, net photochemistry results, as the $Co(CN)_5^{3-}$ product is rapidly intercepted by oxygen to give the Co(III) peroxo complex, $[Co(CN)_5]_2O_2^{6-}$, whose further fate depends on the pH of the solution [157].

A system that, despite its photochemically unreactive behavior, can be included in this section is the binuclear complex $(CN)_5$—Fe—CN—Co(chelate)$^{5-}$ (where "chelate" represents a tetradentate EDTA derivative) studied by Rentzepis and coworkers [158]. In this system, excitation of the Fe(II) → Co(III) IT transition (or localized Co(III) excitation followed by prompt intramolecular electron transfer) generates a substitutionally inert Co(II) chelate fragment. Therefore, contrary to the previously discussed cases, no photoreaction occurs and back electron transfer (of the same type as in Eq. 38) takes place to reform the ground state complex. The interesting aspect of this study is that, by using picosecond absorption techniques, it was possible to distinguish various consecutive steps in the decay of the initially formed IT state. According to the authors, the steps can be assigned to: (i) intersystem crossing from low-spin ($t_{2g}^6e_g$) to high-spin ($t_{2g}^5e_g^2$) Co(II), (ii) back Co(II) → Fe(III) electron transfer to give Co(III) in a triplet ($t_{2g}^5e_g$) excited state, and (iii) relaxation of the excited Co(III) fragment to the ground (t_{2g}^6) state [158]. This study calls the attention on the complexities that may occur in electron transfer processes involving redox couples, such as Co(III)/Co(II), for which the thermodinamically stable oxidized and reduced forms are not related by a simple one-electron transfer step.

3.5 Energy Transfer in Polynuclear Complexes with Cr(III) Luminophoric Units

There are a number of recent studies on intramolecular energy transfer in polynuclear complexes in which Cr(III) complexes have been used as energy accepting fragments. The main reason for this choice lies in the peculiar light emitting properties of Cr(III) complexes.

Octahedral Cr(III) complexes have a quartet ground state $^4A_{2g}$ belonging to the electronic configuration t_{2g}^3. These complexes exhibit relatively weak ligand field bands in the near ultraviolet and visible range. The lowest spin allowed excited state is $^4T_{2g}$ arising from the $t_{2g}^2e_g$ configuration, while the lowest spin-forbidden state is ^2E_g arising from the t_{2g}^3 configuration. The quartet excited state is usually very reactive towards ligand dissociation, lives in the subnanosecond time scale, and undergoes relatively efficient intersystem crossing to the doublet [144]. Because of its intraconfigurational character, on the other hand, the doublet state is a ligand field excited state with rather peculiar properties: (i) its energy depends only slightly on the ligands, through the nephelauxetic effect; (ii) it is essentially unreactive with respect to ligand dissociation; (iii) it has a long (μs to ms) lifetime; (iv) its emission has a very narrow bandshape [144]. These properties make the doublet state easily observable and suggest Cr(III) complexes as convenient light emitting ("luminophoric") fragments to detect intramolecular energy transfer in polynuclear complexes.

3.5.1 Co(III)—Cr(III) and Cr(III)—Cr(III) Binuclear Complexes

The $(CN)_5Co{-}CN{-}Cr(NH_3)_5$ complex has been studied in aqueous solution by Kane-Maguire, et al. [159]. The photochemical and photophysical properties of the mononuclear analogues are as follows [159]. Upon excitation to the first spin-allowed $^1T_{1g}$ state, $Co(CN)_6^{3-}$ gives rise to an efficient photo-aquation reaction ($\Phi = 0.30$), believed to occur in the lowest triplet $^3T_{1g}$ state reached by intersystem crossing. The Co(III) complex is practically non-emitting in room-temperature aqueous solution. On the other hand, the $Cr(NH_3)_6^{3+}$ model compound gives, upon excitation to the lowest spin-allowed state $^4T_{2g}$, both an efficient photoaquation ($\Phi = 0.47$) and emission. The emission is a typical doublet ^2E_g phosphorescence, while the photoaquation is likely a quartet reaction.

The comparison between the absorption spectrum of the $(CN)_5Co{-}CN{-}Cr(NH_3)_5$ binuclear complex and those of the fragments shows that excitation at 313 nm involves almost exclusively the Co(III)-based fragment, whereas light of 436 nm exclusively excites the Cr(III)-based unit. The results obtained upon irradiation at these wavelengths are clear-cut. Irradiation at 313 nm leads to simultaneous bridge cleavage (Eq. 43), release of ammonia (Eq. 44), and Cr(III) doublet emission. Visible excitation at 436 nm, on the other hand,

$$(CN)_5Co{-}CN{-}Cr(NH_3)_5 \xrightarrow[H_2O]{h\nu} Co(CN)_5H_2O^{2-} + Cr(NH_3)_5(CN)^{2+} \tag{43}$$

$$(CN)_5Co{-}CN{-}Cr(NH_3)_5 \xrightarrow[H_2O]{h\nu} (CN)_5Co{-}CN{-}Cr(NH_3)_4H_2O + NH_3 \tag{44}$$

gives rise to release of ammonia (Eq. 44) and Cr(III) doublet emission. The photocleavage quantum yield at 313 nm is only 25% of what expected on the basis of the photoaquation of the mononuclear complex, while the ammonia release and phosphorescence yields are 75% of the corresponding values for 436-nm excitation. This is clear evidence for the occurrence of intramolecular energy transfer from the Co(III)- to the Cr(III)-based unit. As far as the detailed nature of the energy transfer process is concerned, several pathways are energetically allowed in this system, including $^1T_{1g}$-Co(III) or $^3T_{1g}$-Co(III) as energy donors and $^4T_{2g}$-Cr(III) or ^2E_g-Cr(III) as energy acceptors. On the basis of several lines of evidence, among which the coincidence between the extent of quenching of the Co(III) photoreaction and the sensitization of the Cr(III) photoreaction at 313 nm, Kane-Maguire et al. [159] favor a $^3T_{1g}$-Co(III) $\rightarrow$ $^4T_{2g}$-Cr(III) transfer.

The binuclear complex $(CN)(cyclam)Cr{-}CN{-}Cr(CN)_5^-$ has been recently synthesized and studied by Chiorboli et al. [160], with the aim to look for energy transfer between two Cr(III) doublet states. The (CN)Cr(cyclam)—CN— fragment exists as an independent species, $Cr(cyclam)(CN)_2^+$, that in DMF emits a structured phosphorescence at 720 nm with a lifetime of 330 μs [161]. In polynuclear complexes in which this unit is bound to another metal without quenching capability (Sect. 3.5.2), the emission is still expected to be long-lived and easily detectable. A reasonable model for the $—CN—Cr(CN)_5$ fragment is the $Cr(CN)_5NH_3^{2-}$ complex, that is known to emit in DMF solutions at 777 nm with a lifetime of 40 μs [162]. Thus, in the binuclear complex energy transfer from the Cr-cyclam-based fragment to the Cr-cyanide-based one is exothermic by approx. 1000 cm^{-1}.

In the binuclear complex, overlap between the absorption spectra does not permit independent excitation of the two Cr(III) centers. Irrespective of the excitation wavelength, however, the emission expected from the (CN)Cr(cyclam)-CN— fragment is completely quenched, while emission from the $—CN—Cr(CN)_5$ fragment (λ_{max}, 778 nm; τ, 80 μs) is observed in DMF. This strongly suggests that exchange energy transfer from the ^2E_g (O_h) state of the Cr-cyclam-based unit to that of the Cr-cyanide-based one occurs with high efficiency in this system. The excitation spectrum of the 778-nm emission resembles, but is not identical, to the absorption spectrum. This distortion is expected in view of the intersystem crossing efficiency, that is presumably non-unitary and different at the two Cr(III) centers. No risetime in the emission is observed upon laser excitation, indicating that the energy transfer is fast ($k \geq 10^8\ s^{-1}$). This is not unexpected since analogous Cr(III)—Cr(III) bimolecular energy transfer processes, in which the exchange interaction is expected to be much weaker, have rate constants of the order of $10^8\ M^{-1}\ s^{-1}$ [163].

3.5.2 Complexes with Ru(II) Polypyridine Chromophoric Units

While being good light emitters, Cr(III) complexes have weak absorption in the visible because of the ligand-field (d—d) nature of their low-energy transitions. Thus, Cr(III) complexes are ideal luminophoric units to be coupled to strong light absorbing (chromophoric) energy donor units in polynuclear complexes. A number of such Ru(II)—Cr(III) chromophore-luminophore complexes have been recently synthesized and studied.

Bignozzi, et al. [164] have performed a thorough study of the binuclear $(CN)_5Cr—CN—Ru(bpy)_2(CN)^{2-}$ and trinuclear $(CN)_5Cr—CN—Ru(bpy)_2—NC—Cr(CN)_5^{4-}$ complexes. The study was carried out in DMF, where the $Cr(CN)_6^{3-}$ luminophore is known [165] to be a good emitter. The two complexes give rise to very similar results [164]. Visible absorption, that exclusively excites the $—Ru(bpy)_2—$ chromophore, gives rise to efficient emission from the $(CN)_5Cr—CN—$ luminophore, demonstrating the occurrence of exchange energy transfer from the Ru → bpy MLCT triplet state to the Cr doublet state (Fig. 6). The fact that energy transfer does not proceed via the upper quartet state is demonstrated by the lack of the photosubstitutional lability, characteristic of Cr(III) quartet photochemistry [144], in the polynuclear complexes (Fig. 6). The energy transfer processes, that in these systems are exergonic by approx. 3000 cm^{-1}, occur in a subnanosecond time scale. The processes are 100% efficient, leading to a greater efficiency of population of the emitting state relative to that (0.5) of the bare

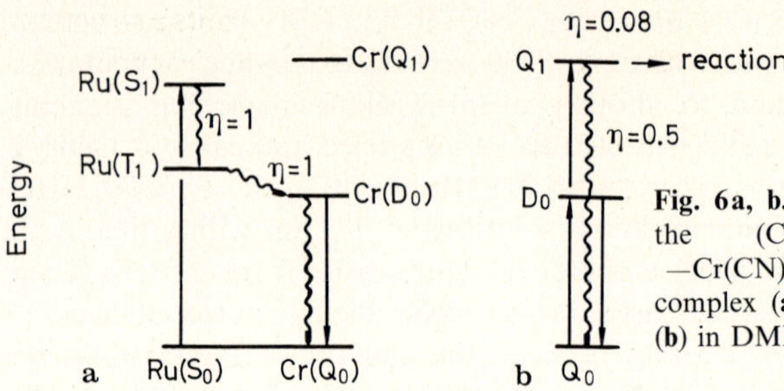

Fig. 6a, b. Photophysical mechanism for the $(CN)_5Cr—CN—Ru(bpy)_2—NC—Cr(CN)_5^{4-}$ chromophore-luminophore complex (**a**) and $Cr(CN)_6^{3-}$ luminophore (**b**) in DMF solutions

luminophore. The behavior of these polynuclear complexes shows some of the ways in which the performance of a luminophore can be improved by covalent coupling to a chromophore: spectral sensitization, antenna effect, enhanced luminescence yields, photoprotection [164].

The absorption spectra of the long-lived doublet state ot these polynuclear complexes exhibit an interesting new type of transition [164], i.e. an intervalence transfer from Ru(II) to *excited* Cr(III) (e.g. Eq. 45)

$$(CN)_5Cr^{III*}—CN—Ru^{II}(bpy)_2(CN)^{2-} \rightarrow$$

$$\rightarrow (CN)_5Cr^{II}—CN—Ru^{III}(bpy)_2(CN)^{2-} \qquad (45)$$

The presence of this IT state above the Cr(III) doublet is responsible, according to the authors [164], for the failure to observe doubly excited Cr(III)*—Ru(II)—Cr(III)* species upon two-photon absorption by the trinuclear complex (as could have been expected on the basis of two successive absorption-energy transfer sequences). Interestingly, the Cr(III)-localized excited states of the bi- and trinuclear complexes give *bimolecular* annihilation processes that are not exhibited by free $Cr(CN)_6^{3-}$. These processes seem to involve intermolecular Ru(II) → Cr(III) electron transfer between two excited polynuclear complexes [164].

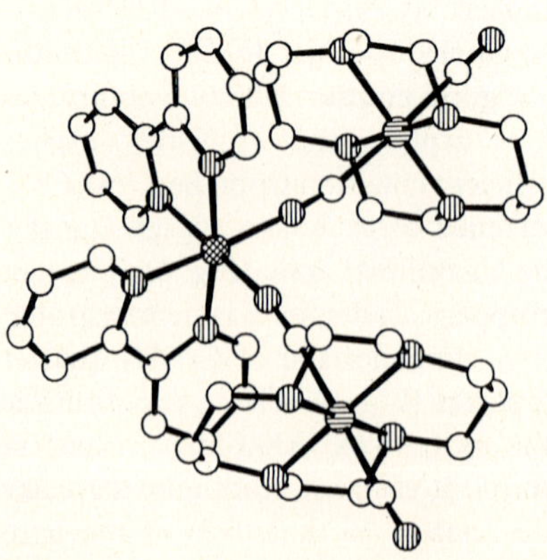

Fig. 7. X-ray structure of the $(NC)Cr(cyclam)—CN—Ru(bpy)_2—NC—Cr(cyclam)(CN)^{4+}$ cation. Ru = ⊗; Cr = ⊜; N = ◍; C = ○

The trinuclear complex (CN)Cr(cyclam)—CN—$Ru(bpy)_2$—NC—Cr(cyclam)-$(CN)^{4+}$ has been recently synthesized by Bignozzi et al. [166]. This constitutes one of the very few cases, among the polynuclear complexes included in this review, for which an X-ray structure is available (Fig. 7). The photophysical behavior of this complex closely parallels that of $(CN)_5Cr$—CN—$Ru(bpy)_2$—NC—$Cr(CN)_5^{4-}$, as far as intramolecular energy transfer, excited-state intervalence transfer, and doublet-doublet annihilation are concerned. Unlike the previous one, this chromophore-luminophore complex emits efficiently ($\Phi = 5.3 \times 10^{-3}$) in aqueous solution [166].

The $(NH_3)_5Cr$—NC—$Ru(bpy)_2$—CN—$Cr(NH_3)_5^{6+}$ complex has been recently studied by Lei and Endicott [167]. This system exhibits efficient energy transfer from the Ru → bpy MLCT triplet to the Cr(III) doublet, as detected by quenching of the MLCT emission and sensitization (77 K) of the Cr(III) phosphorescence. In this system, the energy transfer is estimated to be exergonic by approx. 5000 cm^{-1}. In the same paper, intramolecular energy transfer in analogous complexes containing various Rh(III) ammine fragments instead of the Cr(III) one is also investigated [167].

A specific type of Ru(II)—Cr(III) chromophore-luminophore complex, $Ru(bpy)(CN)_3$—CN—$Cr(NH_3)_5^{+}$, has been recently synthesized and studied by Rampi et al. [168]. The aim was to use second-sphere interactions at the free cyanides of the $Ru(bpy)(CN)_4^{2-}$ chromophore [169] to tune the energy gap between the Ru → bpy MLCT state and the Cr(III) doublet. In this system, efficient intramolecular quenching and sensitization are observed even at the lowest driving forces attainable, i.e. approx. 1000 cm^{-1}.

4 Towards Photonic Molecular Devices

There is currently a growing interest in the possibility of applying the macroscopic concept of "device" to molecules. The idea of "molecular device" is being elaborated in the field of microelectronics, where the possibility of starting from molecules and go "small upward" is seen as the ultimate alternative to the "large downward" approach of lithographic techniques [170–173]. Of course, molecular devices are not futuristic targets but existing operating machines in molecular biology, a field that is indeed deeply permeated by the concept of relationship between molecular structure and function. A molecular device can be defined as an assembly of molecular *components* (i.e. a supramolecular structure) that, because of the specific arrangement of the components in the dimensions of space and energy, is able to perform a *function*. Molecular devices capable of performing light-induced functions (i.e. devices powered by light or capable of elaborating light signals) can be called *photonic molecular devices* (PMD) [34, 174]. General requirements, action mechanisms, machinery, and possible applications of PMDs have been discussed [34, 174]. A number of conceivable simple PMDs are depicted schematically in Fig. 8, where the blocks represent molecular components and the connecting lines suitable chemical links. Several more complex devices can

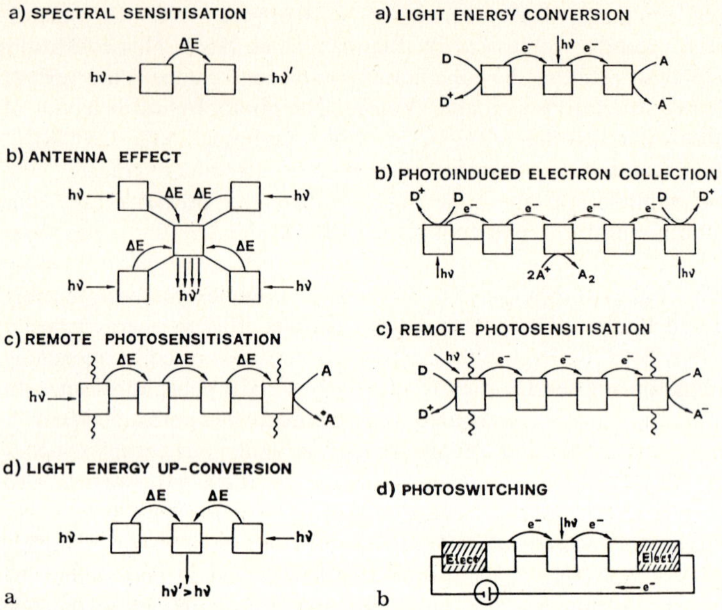

Fig. 8a, b. Block diagrams illustrating the operation of some photonic molecular devices based on (**a**) electronic energy transfer and (**b**) photoinduced electron transfer

be devised, e.g. by combining some of the simple PMDs (e.g. antenna + charge separation, etc.). It can be seen in Fig. 8 that the function is the result of specific sequences of elementary acts performed by the components, that, to that purpose, must be suitably organized in the dimensions of space and energy. The most important elementary acts performed by the components are intramolecular energy transfer (Fig. 8a) and photoinduced electron transfer (Fig. 8b). At present, the design of PMDs is mostly at the stage of an intellectually stimulating problem. There are currently, however, a number of attempts to put these ideas at work, the most outstanding of which are certainly represented by the charge-separating triad systems developed recently in a number of laboratories [123, 124, 175]. It seems likely that the search for useful PMDs will constitute in the near future the main driving force for the study of the photochemistry and photophysics of supramolecular systems.

From the previous sections we have seen that (i) polynuclear complexes are made of components with individual photochemical and photophysical properties and (ii) the behavior of the polynuclear complex is determined by the spatial arrangement and relative energetics of the components. Polynuclear complexes clearly have some of the distinctive features of PMDs. Whether or not a given polynuclear complex is to be considered a PMD depends, of course, on the extent to which its behavior can be considered as a useful light induced function. On very general grounds, it can be pointed out that some of the polynuclear complexes discussed in this article perform interesting functions and could be relevant (as parts of and/or models thereof) to PMDs.

In a number of the polynuclear complexes examined in Sect. 3, e.g. $Os[(dpp)Ru(bpy)_2]_3^{8+}$ (Sect. 3.2.3), $[Ru(bpy)_2(CN)]_2Ru(bpy)_2^{2+}$ (Sect. 3.3.8), and $(CN)Ru(bpy)_2$—NC—$Cr(CN)_5^{2-}$ (Sect. 3.5.2), efficient energy transfer from one or more chromophoric components to a common acceptor component was observed. These systems can be thought as simple PMDs featuring an efficient *antenna* effect. Such polynuclear complexes could serve as energy collecting subunits in various kinds of more complex PMDs for energy conversion. An example of this type of application can be made by considering one of the trinuclear complexes discussed in Sect. 3.3.8. It is known that mononuclear carboxylate derivatives of Ru(II) polypyridine complexes, e.g. $Ru(dc\text{-}bpy)_3^{6-}$, are good sensitizers for performing photocatalytic and photoelectrochemical processes on TiO_2 with visible light [176–178]. Sensitization occurs via adsorption onto the surface of the semiconductor and injection of electrons from the excited sensitizer into the conduction band. As for all dye-sensitized semiconductors, the main drawback of this system is the low optical density achieved in the first, effective adsorbed layer of the dye. A way out of this difficulty is that of using TiO_2 of extremely high surface area [179]. Another possible strategy is that [129] of replacing the mononuclear sensitizer with an *antenna-sensitizer* molecular device. It was seen in Sect. 3.3.8 that the trinuclear complex $(NC)Ru(bpy)_2$—CN—$Ru(dc\text{-}bpy)_2$—NC—$Ru(bpy)_2(CN)^{2-}$ exhibits an efficient "antenna" effect arising from intramolecular energy transfer from the terminal $Ru(bpy)_2$— units to the central —$Ru(dc\text{-}bpy)_2$— component. Since the central component is expected to have good adsorption and photoelectron injection properties on TiO_2, this trinuclear complex seems to be a promising PMD to be used for this purpose. Photocurrent spectra indicate that this trinuclear complex acts indeed as an antenna-sensitizer device on TiO_2 photoelectrodes [129]. It is likely that the use of polynuclear complexes as sensitizers can lead to sizeable improvements in light harvesting efficiency in practical systems, e.g. wet photovoltaic cells, for light energy conversion.

The facile observation of exchange energy transfer in ligand-bridged polynuclear complexes suggests that multi-step energy migration through extended chains or networks of transition metal centers could be feasible, provided that the energy gradient required for each transfer step is small. Future progress in this direction is likely bound to the development of methods for the stepwise synthesis and isolation of oligomeric and polymeric ligand-bridged complexes of known composition.

Polynuclear complexes that feature photoinduced charge separation could in principle be developed following an approach similar to that used in the design of organic charge-separation dyads, triads, tetrads, etc. [123, 124]. Although some success has been obtained with inorganic-organic chromophore-quencher complexes [125, 175], no fully inorganic polynuclear complex with this capability has yet been developed.

An alternative strategy, that may permit the use of transition metal complexes in the construction of PMDs for vectorial charge separation, energy transduction, and photoinduced electron collection is that explored by Meyer [35, 175, 180] and others [181, 182]. In this type of approach, an organic polymeric backbone is used as a pre-formed molecular framework to which metal-containing units (e.g. Ru(II) and Os(II) polypyridine complexes), as well as various organic quenchers, can be covalently attached.

5 Closing Remarks

From the survey of experimental results presented in Sect. 3, the photochemistry of polynuclear complexes appears as a rapidly developing and promising research field.

From a structural point of view, the compounds examined span a remarkable variety of chemical architectures, including binuclear or polynuclear complexes of various complexity, homo and heterometallic systems, stereorigid or flexible systems, one-, two-, and three-dimensional assemblies of metal centers. The synthetic work that underlies the studies in this field, though not discussed at all in this review, is often one of the (if not "the") most important part of this research field. Despite the inherent difficulties, it appears that substantial progress has been made in this direction. In fact, a number of strategies of general synthetic value are emerging, with particular regard to the design of bridging ligands and precursors to metal-containing fragments, and to the assembling reactions (mainly using the method of "complexes as ligands").

From the photochemical and photophysical point of view, the most evident feature is the variety of energy and/or electron transfer pathways accessible with these systems. With respect to other classes of complex molecules, polynuclear complexes appear to be particularly rich systems, in terms of redox sites (metals, bridging ligands, ancillary ligands), fragment-localized excited states (MC, LC, MLCT, MBCT) and intercomponent charge transfer states. Moreover, the energies and redox potentials in these systems are extremely sensitive to changes in metal, bridging ligands and ancillary ligands, and this can be used for tuning purposes. Therefore, there seem to be wide possibilities to use synthetic design to control and orient intramolecular energy and electron transfer in polynuclear complexes.

Of course, the field of the photochemistry and photophysics of polynuclear transition metal complexes is still in its infancy, and much work remains to be done in this area. Further progress in general synthetic methods is required, particularly for complexes of metals other than Ru and Os. Also, some effort to obtain X-ray structures of the polynuclear systems, despite their scarce propensity towards crystallization, should be made. On the mechanistic side, systematic studies on the effect of various parameters (e.g. driving force, nature and length of the bridging group) on the kinetics of intramolecular electron and energy transfer are clearly needed. In fact, although definite proof for the occurrence of intramolecular energy or electron transfer has been obtained in many of the studies discussed in Sect. 3, direct measurements of rate constants are relatively rare. In this respect, the field of polynuclear complexes is still far from the quantitative level reached by the study of organic donor-acceptor systems. Other interesting aspects that seem to deserve future attention in this area are the study of the competition between thermodynamically allowed energy and electron transfer processes and the investigation of multi-step energy or electron transfer along extended chains of metal-containing centers. Finally, the possibility of using polynuclear complexes as building blocks for molecular devices capable of using light energy or processing light signals (photonic molecular devices, Sect. 4) seems to be an exciting direction for future research in this area.

Acknowledgments. The collaboration of Mrs. L. Righetti and Mr. L. Righetti in the preparation of the manuscript is gratefully acknowledged. This work has been supported by the Ministero della Università e della Ricerca Scientifica e Tecnologica and by the Consiglio Nazionale delle Ricerche.

6 Appendix

6.1 Ligand Abbreviations

Abbreviation	Formula Number
bibpy-2	23 with R = $(CH_2)_2$
bibpy-3	23 with R = $(CH_2)_3$
bibpy-5	23 with R = $(CH_2)_5$
bibpy-12	23 with R = $(CH_2)_{12}$
bibpy-3′	23 with R = $CH_2{-}CHOH{-}CH_2$
bibpy-8′	23 with R = $CH_2{-}C_6H_4{-}CH_2$
bidpq	18
biq	5
bpa	9
bpe	10
bpm	11
bptH	27
bpy	3 with R = H
4,4′-bpy	8
bpz	12
cyclam	26
dc-bpy	3 with R = COO^-
dec-bpy	3 with R = $COOC_2H_5$
dpp	14
2,5-dpp	13
dppe	24
dpq	16
dpte	25
HAT	21
Me_2-bpy	3 with R = CH_3
Mepz	1 with R = CH_3

phen	6
ppz	15
py	2
pz	1 with R = H
QP	20
R_2-bpy	3
tpp	19
tppq	17
tpy	7
tribpy	22

6.2 Structural Formulae of the Ligands

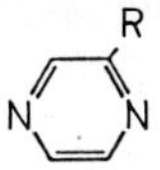

1

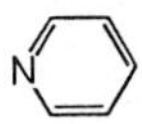

2

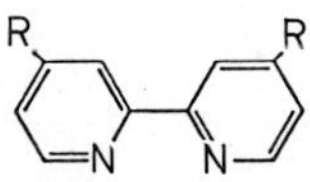

3

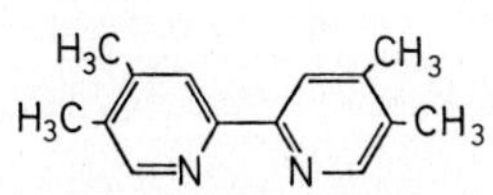

4

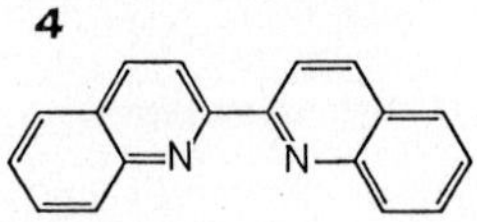

5

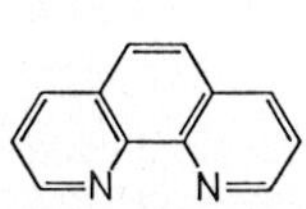

6

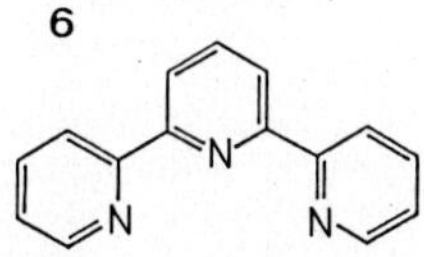

7

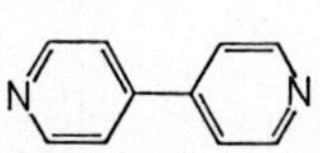

8

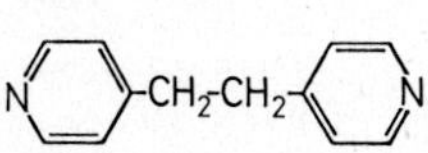

9

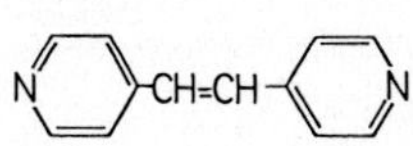

10

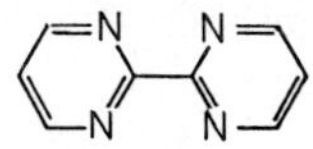

11

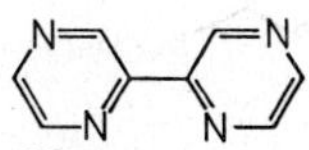

12

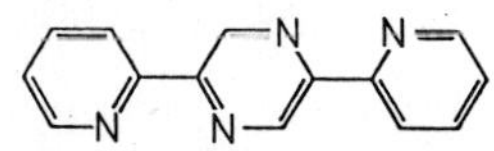

13

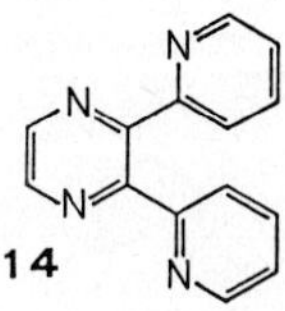

14

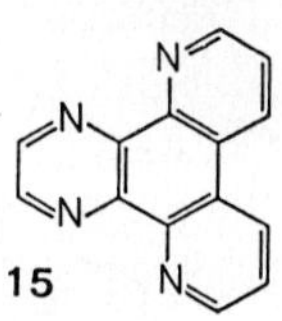

15

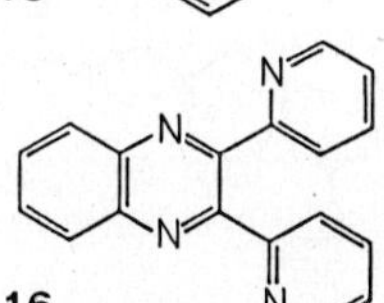

16

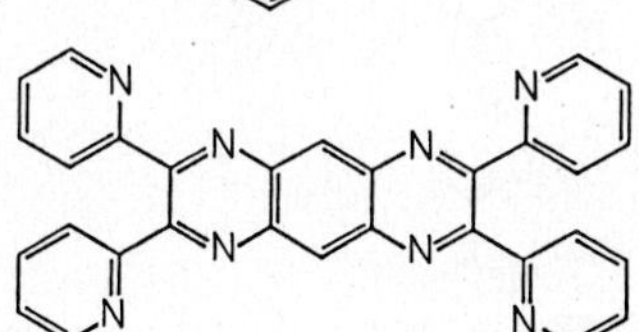

17

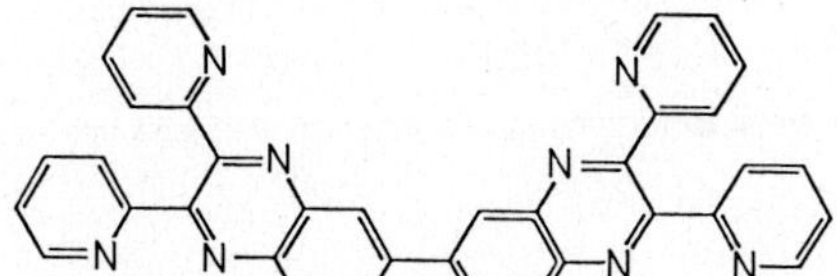

18

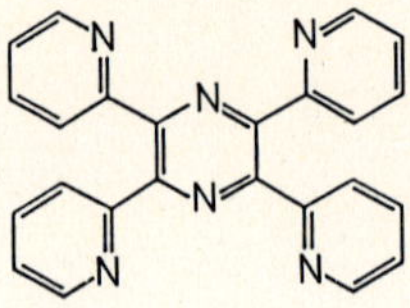

19

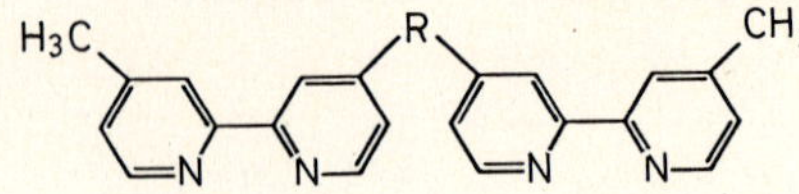

23

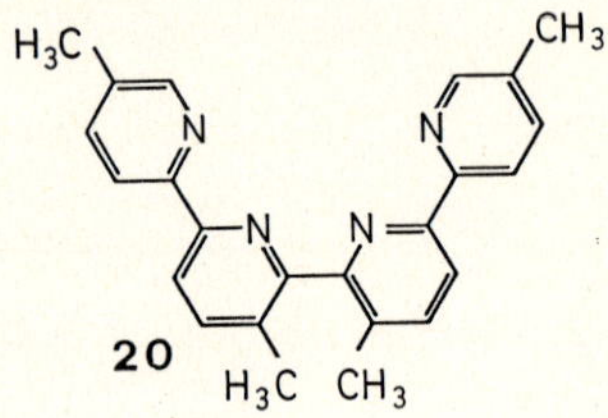

20

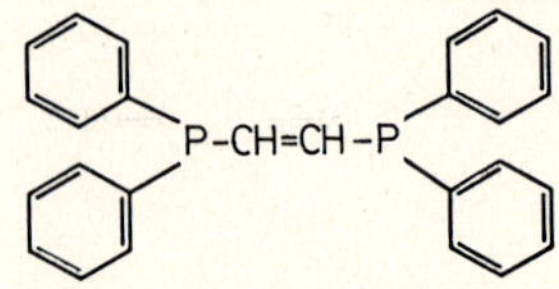

24

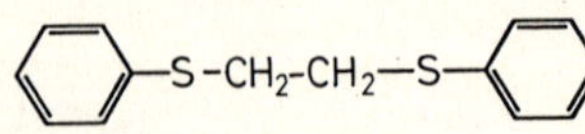

25

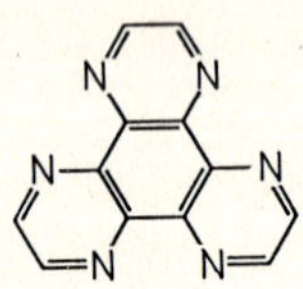

21

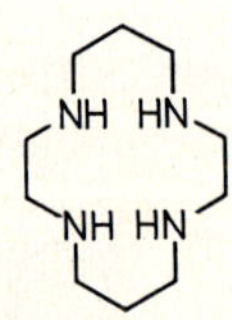

26

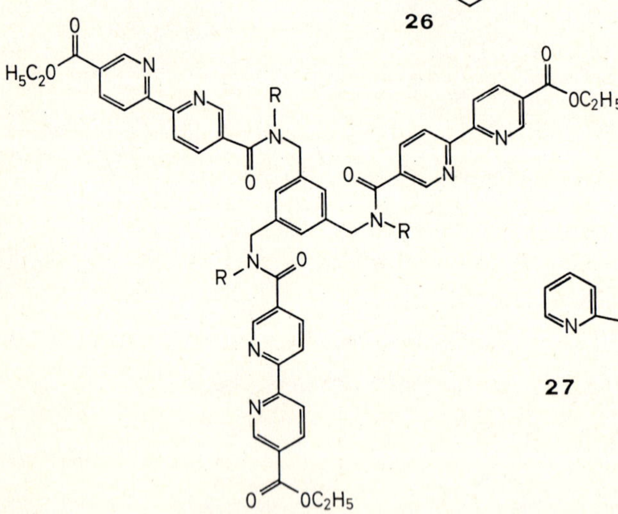

22

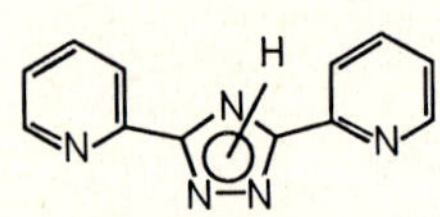

27

7 References

1. Plane RA, Hunt JP (1957) J. Am. Chem. Soc. 79: 3343
2. Schläfer HL (1957) Z. Physik. Chem. Frankfurt 11: 65
3. Adamson AW, Sporer AH (1958) J. Am. Chem. Soc. 80: 3865
4. Carassiti W, Claudi M (1959) Ann. Chim. Rome 49: 1697
5. Balzani V, Carassiti V (1970) Photochemistry of coordination compounds, Academic, London
6. Adamson AW, Fleischauer PD (eds) (1975) Concepts of inorganic photochemistry. Wiley, New York
7. Geoffroy GL, Wrighton MS (eds) (1973) Organometallic photochemistry, Academic, New York
8. Special issue of J. Chem. Educ. (1983) 60: 785–887
9. Special issue of Coord. Chem. Rev. (1985) 64: 1–385
10. Lever ABP (ed) (1986) Excited states and reactive intermediates. ACS Symposium Series n° 307 ACS, Washington
11. Yersin H, Vogler A (eds) (1987) Photochemistry and photophysics of coordination compounds, Springer, Berlin Heidelberg New York
12. Ferraudi GJ (ed) (1988) Elements of inorganic photochemistry. Wiley, New York
13. Special issue of Coord. Chem. Rev. (1990) 97: 1–326
14. Demas JN, Adamson AW (1971) J. Am. Chem. Soc. 93: 1800
15. Balzani V, Moggi L, Manfrin MF, Bolletta F, Laurence GS (1975) Coord. Chem. Rev. 15: 321
16. Balzani V, Bolletta F, Gandolfi MT, Maestri M (1978) Top. Curr. Chem. 75: 1
17. Sutin N (1983) J. Chem. Educ. 60: 809
18. Scandola F, Balzani V (1983) J. Chem. Educ. 60: 814
19. Hoffman MZ, Bolletta F, Moggi L, Hug GL (1989) J. Phys. Chem. Ref. Data 18: 219
20. Meyer TJ (1978) Acc. Chem. Res. 11: 94
21. Kalyanasundaram K (1982) Coord. Chem. Rev. 46: 159
22. Watts J (1983) J. Chem. Educ. 60: 834
23. Seddon EA, Seddon KR (eds) (1984) The chemistry of ruthenium, Elsevier, Amsterdam, chap. 15
24. Juris A, Balzani V, Barigelletti F, Campagna S, Belser P, von Zelewsky A (1988) Coord. Chem. Rev. 84: 85
25. Bolton JR (ed) (1977) Solar power and fuels, Academic, New York
26. Connolly JS (ed) (1981) Photochemical conversion and storage of solar energy. Academic Press, London
27. Harriman A, West MA (eds) (1983) Photogeneration of hydrogen, Academic, New York
28. Grätzel M (ed) (1983) Energy resources through photochemistry and catalysis, Academic, New York
29. Norris JR, Meisel D (eds) (1989) Photochemical energy conversion, Elsevier, New York
30. Lehn J-M (1988) Angew. Chem. Int. Ed. Engl. 27: 89
31. Cram DJ (1988) Angew. Chem. Int. Ed. Engl. 27: 1009
32. Kohnke FH, Mathias JP, Stoddart JF (1989) Angew. Chem. Int. Ed. Engl. Adv. Mater. 28: 1103
33. Balzani V (ed) (1987) Supramolecular photochemistry, Reidel, Dordrecht
34. Balzani V, Scandola F (submitted for publ.)
35. See, e.g., Younathan JN, McClanahan SF, Meyer TJ (1989) Macromolecules 22: 1048 and references therein
36. See, e.g., Meade TJ, Gray HB, Wimkler JR (1989) J. Am. Chem. Soc. 111: 4353 and references therein
37. See, e.g., Menon RK, Brown TL (1989) Inorg. Chem. 28: 1370 and references therein
38. See, e.g., Davila J, Harriman A, Milgrom LR (1987) Chem. Phys. Lett. 136: 427 and references therein
39. See, e.g., Gubelmann M, Harriman A, Lehn J-M, Sessler JL (1988) J. Chem. Soc. Chem. Commun. 77 and references therein

40. Connolly JS, Bolton JR (1988) In: Fox MA, Chanon M (eds) Photoinduced electron transfer Part D, Elsevier, Amsterdam, p 303
41. See, e.g., Chardon-Noblat S, Sauvage JP, Mathis P (1989) Angew. Chem. Int. Ed. Engl. 28: 593 and references therein
42. Marcus RA (1964) Annu. Rev. Phys. Chem 15: 155
43. Hush NS (1968) Electrochim. Acta 13: 1005
44. Sutin N (1979) In: Eichorn GL (ed) Inorganic biochemistry, Elsevier, New York, p 611
45. Sutin N (1983) Prog. Inorg. Chem. 30: 441
46. Marcus RA, Sutin N (1985) Biochim. Biophys. Acta 811: 265
47. Ulstrup J (1979) Charge transfer in condensed media, Springer, Berlin Heidelberg New York, and references therein
48. Rehm D, Weller A (1970) Isr. J. Chem. 8: 259
49. Indelli MT, Ballardini R, Scandola F (1984) J. Phys. Chem. 88: 2547 and references therein
50. Closs GL, Miller JR (1988) Science 240: 440
51. Gould IR, Moser JE, Armitage B, Farid S (1989) J. Am. Chem. Soc. 111: 1917
52. Mayoh B, Day P (1972) J. Am. Chem. Soc. 94: 2885
53. Richardson DE, Taube H (1983) J. Am. Chem. Soc. 105: 40
54. Endicott JN (1988) Acc. Chem. Res. 21: 59
55. Creutz C (1983) Prog. Inorg. Chem. 30: 1
56. Oevering H, Paddon-Row MN, Heppener M, Oliver AM, Cotsaris E, Verhoeven JW, Hush NS (1987) J. Am. Chem. Soc. 109: 3258
57. Turro NJ (1978) Modern molecular photochemistry, Benjamin, Menlo Park, Ca
58. Lamola AA (1969) in: Lamola AA, Turro NJ (eds) Energy transfer and organic photochemistry, Interscience, New York, p 17
59. Witt H (1987) Nouv. J. Chim. 11: 91
60. Crosby GA (1983) J. Chem. Educ. 60: 791
61. Closs GL, Piotrowiak P, MacInnis JM, Fleming GR (1988) J. Am. Chem. Soc. 110: 2652
62. Closs GL, Johnson MD, Miller JR, Piotrowiak P (1989) J. Am. Chem. Soc. 111: 3751 and references therein
63. Robin MB, Day P (1967) Adv. Inorg. Chem. Radiochem. 10: 247
64. Hush NS (1967) Prog. Inorg. Chem. 8: 391
65. Balzani V, Scandola F (1988) In: Fox MA, Chanon M (eds) Photoinduced electron transfer Part D, Elsevier, Amsterdam, p 148
66. Wacholtz WF, Auerbach RA, Schmehl RH (1987) Inorg. Chem. 26: 2989
67. Furue M, Kuroda N, Sano S (1988) J. Macromol. Sci. Chem. A25: 1263
68. Furue M, Kuroda N, Nozakura S (1986) Chem. Lett. 1209
69. Sahai R, Baucom DA, Rillema DP (1986) Inorg. Chem. 25: 2843
70. De Cola L, Belser P, Ebmeyer F, Barigelletti F, Vögtle F, von Zelewsky A, Balzani V (1990) Inorg. Chem. 29: 495
71. Furue M, Kinoshita S, Kushida T (1987) Chem. Lett. 2355
72. Schmehl RH, Auerbach RA, Wacholtz WF, Elliott CM, Freitag RA, Merkert JW (1986) Inorg. Chem. 25: 2440
73. Schmehl RH, Auerbach RA, Wacholtz WF (1988) J. Phys. Chem. 92: 6202
74. Petersen JD, Murphy WR jr, Sahai R, Brewer KJ, Ruminski RR (1985) Coord. Chem. Rev. 64: 261
75. Ruminski R, Kiplinger J, Cockroft T, Chase C (1989) Inorg. Chem. 28: 370
76. Haga M, Matsumura-Inoue T, Yamabe S (1987) Inorg. Chem. 26: 4148
77. Ernst S, Kasack V, Kaim W (1988) Inorg. Chem. 27: 1146
78. Hunziker M, Ludi A (1977) J. Am. Chem. Soc. 99: 7370
79. Dose EV, Wilson LJ (1978) Inorg. Chem. 17: 2660
80. Braunstein CH, Baker AD, Strekas TC, Gafney HD (1984) Inorg. Chem. 23: 857
81. Fuchs Y, Lofters S, Dieter T, Shi W, Morgan R, Strekas TC, Gafney HD, Baker AD (1987) J. Am. Chem. Soc. 109: 2691
82. Petersen JD (1987) In: Yersin H, Vogler A (eds) Photochemistry and photophysics of coordination compounds, Springer, Berlin Heidelberg New York, p 147

83. Petersen JD (1987) In: Balzani V (ed) Supramolecular photochemistry. Reidel, Dordrecht, p 135
84. Kalyanasundaram K, Nazeeruddin MdK (1990) Inorg. Chem. 29: 1880
85. Hage R, Dijkhuis AHJ, Haasnoot JG, Prins R, Reedijk J (1988) Inorg. Chem. 27: 2185
86. Hage R, Haasnoot JG, Stufkens DJ, Snoeck TL, Vos JG, Reedijk J (1989) Inorg. Chem 28: 1413
87. Barigelletti F, DeCola L, Balzani V, Hage R, Haasnoot JG, Reedijk J, Vos JG (1989) Inorg. Chem. 28: 4344
88. Sahai R, Rillema DP (1986) Inorg. Chim. Acta 118: L35
89. Kalyanasundaram K, Nazeeruddin MdK (1989) Chem. Phys. Lett. 158: 45
90. Lin C-T, Böttcher W, Chou M, Creutz C, Sutin N (1976) J. Am. Chem. Soc. 98: 6536
91. Creutz C, Sutin N (1976) Inorg. Chem. 15: 496
92. Masschelein A, Kirsch-De Mesmaeker A, Verhoeven C, Nasielski-Hinkens R (1987) Inorg. Chim. Acta 129: L13
93. Kirsch-De Mesmaeker A, Jacquet L, Masschelein A, Vanhecke F, Heremans K (1989) Inorg. Chem. 28: 2465
94. Rillema DP, Callahan RW, Mack KB (1982) Inorg. Chem. 21: 2589
95. Sahai R, Rillema DP (1986) J. Chem. Soc. Chem. Commun. 1133
96. Campagna S, Denti GF, Sabatino L, Serroni S, Ciano M, Balzani V (1989) Gazz. Chim. It. 119: 415
97. Campagna S, Denti GF, Sabatino L, Serroni S, Ciano M, Balzani V (1989) J. Chem. Soc. Chem. Commun. 1500
98. Murphy WR Jr, Brewer KJ, Gettliffe G, Petersen JD (1989) Inorg. Chem. 28: 81
99. Sahai R, Morgan L, Rillema DP (1988) Inorg. Chem. 27: 3495
100. Rillema DP, Mack KB (1982) Inorg. Chem. 21: 3849
101. Toma HE, Auburn PR, Dodsworth E, Golovin MN, Lever ABP (1987) Inorg. Chem. 26: 4257
102. Von Kameke A, Tom GM, Taube H (1978) Inorg. Chem. 17: 1790
103. Toma HE, Lever ABP (1986) Inorg. Chem. 25: 176
104. Zulu MM, Lees AJ (1988) Inorg. Chem. 27: 1139; Zulu MM, Lees AJ (1988) Inorg. Chem. 27: 3325
105. Zulu MM, Lees AJ (1989) Inorg. Chem. 28: 85
106. Campagna S, Denti GF, De Rosa G, Sabatino L, Ciano M, Balzani V (1989) Inorg. Chem. 28: 2565
107. Shoup M, Hall B, Ruminski RR (1988) Inorg. Chem. 27: 200
108. Vogler A, Kisslinger J (1986) Inorg. Chim. Acta 115: 193
109. Juris A, Campagna S, Bidd I, Lehn JM, Ziessel R (1988) Inorg. Chem. 27: 4007
110. Sahai R, Rillema DP, Shaver R, Wallendael SV, Jackman DC, Boldaji M (1989) Inorg. Chem. 28: 1022
111. Creutz C, Kroger P, Matsubara T, Netzel TL, Sutin N (1979) J. Am. Chem. Soc. 101: 5442
112. Durante VA, Ford PC (1975) J. Am. Chem. Soc. 97: 6898
113. Curtis JC, Bernstein JS, Meyer TJ (1985) Inorg. Chem. 24: 385
114. Vogler A, Kunkely H (1981) J. Am. Chem. Soc. 103: 1559
115. Heath GA, Yellowlees LJ, Braterman PS (1982) Chem. Phys. Lett. 92: 646
116. Perkins TA, Pourreau DB, Netzel TL, Schanze KS (1989) J. Phys. Chem. 93: 4511
117. Winkler JR, Netzel TL, Creutz C, Sutin N (1987) J. Am. Chem. Soc. 109: 2381
118. Schanze KS, Meyer TJ (1985) Inorg. Chem. 24: 2123
119. Schanze KS, Neyhart GA, Meyer TJ (1986) J. Phys. Chem. 90: 2182
120. Tapolsky G, Duesing R, Meyer TJ (1989) J. Phys. Chem. 93: 3885
121. Bignozzi CA, Roffia S, Scandola F (1985) J. Am. Chem. Soc. 107: 1644; Bignozzi CA, Paradisi C, Roffia S, Scandola F (1988) Inorg. Chem. 27: 408
122. Scandola F (1989) In: Morris JR, Meisel D (eds) Photochemical energy conversion, Elsevier, New York, p 60
123. Wasielewski MR, Niemczyk MP, Svec WA, Pewitt EB (1985) J. Am. Chem. Soc. 107: 5562
124. Gust D, Moore TA (1989) Science 244: 35

125. Danielson E, Elliott CM, Merkert JW, Meyer TJ (1987) J. Am. Chem. Soc. 109: 2519
126. Katz NE, Creutz C, Sutin N (1988) Inorg. Chem. 27: 1687
127. Bignozzi CA, Roffia S, Chiorboli C, Davila J, Indelli MT, Scandola F (1989) Inorg. Chem. 28: 4350
128. Meyer TJ (1986) Pure Appl. Chem. 58: 1193
129. Amadelli R, Argazzi R, Bignozzi CA, Scandola F J. Am. Chem. Soc. in press
130. Oevering H, Verhoeven JW, Paddon-Row MN, Cotsaris E, Hush NS (1988) Chem. Phys. Lett. 143: 488
131. Wasielewski MR (1988) In: Fox MA, Chanon M (eds) Photoinduced electron transfer Part A, Elsevier, Amsterdam, p 161
132. Fox LS, Marshall JL, Gray HB (1987) J. Am. Chem. Soc. 109: 6901
133. Chen P, Westmoreland TD, Danielson E, Schanze KS, Anthon D, Neveux Jr PE, Meyer TJ (1987) Inorg. Chem. 26: 1116
134. Schanze KS, Sauer K (1988) J. Am. Chem. Soc. 110: 1180
135. Cooley LF, Headford CEL, Elliott CM, Kelley DF (1988) J. Am. Chem. Soc. 110: 6673
136. McMahon RJ, Forcé RK, Patterson HH, Wrighton MS (1988) J. Am. Chem. Soc. 110: 2670
137. Chen P, Curry M, Meyer TJ (1989) Inorg. Chem. 28: 2271
138. Boyde S, Strouse GF, Jones WE Jr, Meyer TJ (1989) J. Am. Chem. Soc. 111: 7448
139. Franco C, McLendon G (1984) Inorg. Chem. 23: 2370
140. Thorn DL, Fultz WC (1989) J. Phys. Chem. 93: 1234
141. Vogler A, Osman AH, Kunkely H (1985) Coord. Chem. Rev. 64: 159
142. Hennig H, Rehorex A, Rehorex D, Thomas P (1984) Inorg. Chim. Acta 86: 41
143. Nishizawa M, Ford PC (1981) Inorg. Chem. 20: 2016
144. Zinato E (1975) In: Adamson AW, Fleischauer PD (eds) Concepts of inorganic photochemistry, Wiley, New York, p 143
145. Gelroth JA, Figard JE, Petersen JD (1979) J. Am. Chem. Soc. 101: 3649
146. Moore KJ, Lee L, Figard JE, Gelroth JA, Stinson AJ, Wohlers HD, Petersen JD (1983) J. Am. Chem. Soc. 105: 2274
147. Figard JE, Petersen JD (1978) Inorg. Chem. 17: 1059
148. Ford PC, Hintze RE, Petersen JD (1975) In: Adamson AW, Fleischauer PD (eds) Concepts of inorganic photochemistry, Wiley, New York, p 203
149. MacQueen DB, Petersen JD (submitted for publ.)
150. Malin JM, Ryan DA, O'Halloran TV (1978) J. Am. Chem. Soc. 100: 2097
151. Buhks E, Bixon M, Jortner J, Navon G (1981) J. Phys. Chem. 85: 3759
152. Norton KA Jr, Hurst JK (1982) J. Am. Chem. Soc. 104: 5960
153. Farr JK, Hulett LG, Lane RH, Hurst JK (1975) J. Am. Chem. Soc. 97: 2654
154. Vogler A, Kunkely H (1975) Ber. Bunsen-Ges. Phys. Chem. 79: 83
155. Vogler A, Osman AH, Kunkely H (1987) Inorg. Chem. 26: 2337
156. Bagger S, Stoltze P (1981) Acta Chem. Scand. A35: 509
157. Vogler A, Kunkely H (1975) Ber. Bunsen-Ges. Phys. Chem. 79: 301
158. Reagor BT, Kelley DF, Huchital DH, Rentzepis PM (1982) J. Am. Chem. Soc. 104: 7400
159. Kane-Maguire NAP, Allen MM, Vaught JM, Hallock JS, Heatherington AL (1983) Inorg. Chem. 22: 3851
160. Chiorboli C, Indelli MT, Bignozzi CA, Scandola F (manuscript in preparation)
161. Miller DB, Miller PK, Kane-Maguire NAP (1983) Inorg. Chem. 22: 3831
162. Riccieri P, Zinato E (manuscript in preparation)
163. Endicott JF, Lessard RB, Lei Y, Ryu CK (1987) in: Balzani V (ed) Supramolecular Photochemistry. Reidel, Dordrecht, p 167
164. Bignozzi CA, Indelli MT, Scandola F (1989) J. Am. Chem. Soc. 111: 5192
165. Wasgestian HF (1972) J. Phys. Chem. 76: 1947
166. Bignozzi CA, Chiorboli C, Indelli MT, Scandola F (manuscript in preparation)
167. Lei Y, Endicott JF (submitted for publ.)
168. Rampi MA, Checchi L, Scandola F (manuscript in preparation)
169. Scandola F, Indelli MT (1988) Pure Appl. Chem. 60: 973
170. Carter FL (ed) (1987) Molecular electronic devices II, Dekker, New York

171. Haddon RC, Lamola A (1985) Proc. Natl. Acad. Sci. USA 82: 1874
172. Aviram A (1988) J. Am. Chem. Soc. 110: 5687
173. Hopfield JJ, Onuchic JN, Beratan DN (1989) J. Phys. Chem. 93: 6350
174. Balzani V, Moggi L, Scandola F (1987) In: Balzani V (ed) Supramolecular photochemistry, Reidel, Dordrecht, p 1
175. Meyer TJ (1989) Accounts Chem. Res. 22: 163
176. Desilvestro J, Grätzel M, Kavan L, Moser J (1985) J. Am. Chem. Soc. 107: 2988
177. Furlong DN, Wells D, Sasse WHF (1986) J. Phys. Chem. 90: 1107
178. Vlachopoulos N, Liska P, Augustynski J, Grätzel M (1988) J. Am. Chem. Soc. 110: 1216
179. Grätzel M (1989) in: Norris JR, Meisel D (eds) Photochemical energy conversion, Elsevier, New York, p 284
180. Strouse GF, Worl LA, Younathan JN, Meyer TJ (1989) J. Am. Chem. Soc. 111: 9101
181. Kaneko M, Nakamura H (1987) Makromol. Chem. 188: 2011
182. Ennis PM, Kelly JM (1989) J. Phys. Chem. 93: 5735

Photoinduced Electron Transfer in Ion Pairs

Roland Billing, Detlef Rehorek and Horst Hennig

Sektion Chemie der Karl-Marx-Universität Leipzig, Talstraße 35, DDR-7010, Leipzig

Table of Contents

Abbreviations

$acac^-$	acetylacetonate (pentane-2,4-dionate)
AMS^-	anilinomethane sulfonate
bpy	2,2′-bipyridine
Bu	*n*-butyl
BV^{2+}	1,1′-dibenzyl-4,4′-bipyridinediium
$CDTA^{4-}$	trans-1,2-cyclohexane diaminetetraacetate
cp	cyclopentadienyl
$DMBPY^{2+}$	1,1′-dimethyl-2,2′-bipyridinediium
dmp	2,9-dimethyl-1,10-phenanthroline
dtc^-	diethyldithiocarbamate
$EDTA^{4-}$	ethylenediaminetatraacetate
en	1,2-diaminoethane
EY^{2-}	eosin anion
htc^-	bis(2-hydroxyethyl)dithiocarbamate
ICC^+	di-(*n*-butyl)tetramethylindocarbocyanine
IDA^{2-}	iminodiacetate
ImH	imidazole
KOS^+	4-carboxymethyl-1-ethylpyridinium
$Lucigenin^{2+}$	10,10′-dimethyl-9,9′-biacridinediium
MB^+	methylene blue
Me	methyl
mnt^{2-}	maleonitrile-1,2-dithiolate
MPZ^+	1-methylpyrazinium
MQX^+	1-methylquinoxalinium
MV^{2+}	1,1′-dimethyl-4,4′-bipyridinediium
$NSPY^+$	4-nitrostyryl-4′-pyridinium
$PDTA^{4-}$	propanediaminetetraacetate
Ph	phenyl
phen	1,10-phenanthroline
$PMAA^{n-}$	poly(methacrylate)
py	pyridine
$Quin^+$	quinolinium
RB^{2-}	rose bengale dianion
sep	1,3,6,8,10,13,16,19-octaazabicyclo[6.6.6.]eicosane
tacn	1,4,7-triazacyclononane
$TFPB^-$	tetrakis[3,5-bis(trifluoromethyl)phenyl]borate

tim — 2,3,9,10-tetramethyl-1,4,8,11-tetra-azacyclotetradeca-1,3,8,10-tetraene
TPP^+ — 1,3,5-triphenylpyrylium
tmen — 1,2-bis(dimethylamino)ethane
$TPPS^{4-}$ — *meso*-tetrakis(4-sulfonatophenyl)porphyrin
$TROP^+$ — tropylium
$TSPC^{4-}$ — tetrakis(4-sulfonatophenyl)phthalocyanin
$TTAP^{4+}$ — *meso*-tetrakis(trimethylammoniophenyl)porphyrin
(2.2.1.) — 4,7,13,16,21-pentaoxo-1,10-diazabicyclo[8.8.5]-tricosane

Electrostatic attraction by oppositely charged ions leads to ion pairing. The influence of solvent polarity, the size and the charge of the ions as well as the electron-pair donor and acceptor strength on the equilibrium constant of ion-pair formation is discussed.

Although ion-pairing leads to only loosely bound species, a marked influence on both spectroscopic and photochemical properties of the participating ions has been observed.

The conditions to be met for the appearance of additional IPCT absorption bands as well as the influence of the solvent and the redox potentials of the ions on the energy of IPCT states will be discussed in some detail.

Light absorption by ion pairs is associated with electron transfer, either directly or via subsequent thermal processes, which leads to permanent formation of photoproducts. It is shown that other fast processes have to compete with back electron transfer in order to achieve high product yields.

Mechanistic aspects of the formation of permanent photoproducts are discussed with respect to ion pairs of aryldiazonium and diaryliodonium cations, cobalt(III) complexes, tetraarylborates and carboxylates.

1 Introduction

Photoinduced electron-transfer reactions have been the subject of intense studies for the last few decades [1]. It is the essence of these reactions that either directly by irradiation with light or by subsequent thermal reactions, an electron is transferred from the donor to the acceptor. Electron transfer may lead to a dramatic alteration of the chemical behavior, e.g. redox properties, kinetic stability, catalytic activity, acid-base properties, which provide the basis for the formation of permanent photoproducts. These processes may be used e.g. for the design of unconventional information recording materials, for solar energy conversion, and for the generation of catalysts in organic synthesis [2, 3].

Since the three functional components of photoinduced electron transfer, i.e. donor, acceptor and light-absorbing chromophor, are very often not contained in one molecular entity, the efficiency of electron transfer processes is markedly reduced by diffusion control of the formation of encounter complexes [4]. This disadvantage of bimolecular reactions can be overcome by creating conditions under which all functional components are bound together in one stable complex (supramolecule) prior to light absorption [5].

With respect to metal complexes, this idea can be realized by interactions with additional ligands. This phenomenon, which is also known as second-sphere coordination, outer-sphere coordination and supracomplex formation and which was first mentioned by Werner [6] in 1913, has become increasingly interesting during the recent years [7–11].

Ion pairing between oppositely charged ions renders one of the various possibilities to bring about second-sphere interactions. As compared with interactions within the first coordination sphere, these interactions are weak and lead to relatively labile complexes which are distinguished from collisional encounter complexes by a well-defined array of ligands in the second coordination sphere [10]. Due to the weakness of those interactions, the chemical identity of the coordination compound remains essentially the same. On the other hand, the formation of ion pairs leads to new electronic states which may effect the spectroscopic, photophysical, and photochemical behavior of the coordination compound [7]. Ion pairing thus gives a possibility of varying photochemical properties of charged coordination compounds while the first coordination sphere remains intact. This also holds true for ion pairs not containing metal complexes.

In this article, we shall concentrate on ion pairs formed by association of oppositely charged, individually existing ions. However, ion pairs formed by association of negatively charged metal complexes with metal cations, e.g. Prussian Blue analogs, will be omitted. For a comprehensive review on these intervalence charge-transfer complexes the reader is referred to a series of excellent monographs and review articles, e.g. [12, 13].

2 Formation of Ion Pairs

2.1 Experimental Methods for the Determination of Ion-Pair Formation Constants

The ion pair formation constant K_{IP} is given by the equilibrium (1) between free and ion-paired cations $[A^+]$ and anions $[B^-]$, respectively

$$A^+ + B^- \rightleftharpoons \underset{\text{ion pair}}{A^+;B^-} \quad (1)$$

$$K_{IP} = [A^+;B^-]/([A^+][B^-]) \quad (2)$$

Ion pairing leads to an alteration of a large variety of physical and chemical properties. Therefore, both spectroscopic and nonspectroscopic methods can be used for the determination of the ion pair formation constant K_{IP} (Table 1). However, for ion pairs exhibiting additional ion-pair charge-transfer (IPCT) absorptions, the measurement of the increase of the optical density and treatment of the deviation from the Lambert-Beer law by methods similar to those given by Benesi and Hildebrand [14] and Drago and Rose [15], respectively, belong to the most popular ones.

Table 1. Molecular properties of ion pairs and methods used for the determination of ion pair formation constants

Molecular property	Method	Ref.
specific conductivity	conductometry	[16]
charge, volume	chromatography	[17]
relaxation time	NMR spectroscopy	[18]
dipolar momentum	IR spectroscopy	[19]
induced circular dichroism	OCD measurement	[20]
absorption of ultrasound	ultrasound spectroscopy	[21]
electronic energy	UV/VIS spectroscopy	[22]
spin density	EPR spectroscopy	[23]
luminescence lifetime	emission spectroscopy	[24]
light sensitivity	flash photolysis	[25]

2.2 Electrostatic Theory

Assuming that the participating ions may be treated as rigid spheres having electric chargers z_+ and z_- in their centers and that a purely electrostatic interaction takes place in an unstructured solvent of static dielectric constant D_s, the ion pair formation constant K_{IP} may be calculated according to Eigen and Fuoss [26–28], Eq. (3)

$$K_{IP} = \frac{4\pi N d^3}{3000} \exp[-U(d)/RT] \tag{3}$$

where

$$U(d) = \frac{z_+ z_- e^2}{D_s d(1 + \varkappa d)}$$

and

$$\varkappa = (8\pi N e^2 I/1000 D_s RT)^{1/2}$$

N — Avogadro's number,
d — contact distance, i.e. sum of radii,
z_+, z_- — charge of the cation and anion, respectively,
e — electronic charge,
D_s — static dielectric constant,
I — ionic strength

The Eigen-Fuoss equation (3) generally leads to fairly good agreement between experimental and calculated ion pair formation constants indicating that the interactions between ions are indeed basically of an electrostatic nature. Considerable deviation between experimental and calculated K_{IP} is observed when additional interactions between the ions occur, e.g. hydrogen bonds between interacting cyanometallates and cationic cobalt(III) ammine complexes [28].

From Eq. (3) it follows that the extent of ion pairing may be increased by choosing nonpolar solvents and low ionic strength.

2.3 Electron-Pair Donor-Acceptor Interactions

For a detailed, though qualitative, treatment of bonding interactions in addition to electrostatic interactions, the concept of donors and acceptors proposed by Gutmann [29, 30] offers a satisfactory basis. In the framework of this concept, ion pairing is regarded as a competitional process between the addition of a solvent molecule and the addition of a counterion at a suitable site of the second coordination sphere of the complex. The formation of ion pairs should be particularly favored if the cation is a better σ-electron-pair acceptor and the anion is a better σ-electron-pair donor than the solvent.

For the determination of the corresponding donor and acceptor numbers of the ions, Soukup [31] has proposed the solvatochromic systems $[Cu(acac)(tmen)]^+$ and $[Fe(phen)_2(CN)_2]$. The σ-electron-pair acceptor and σ-electron-pair donor properties of a variety of ions together with those of most commonly used solvents are summarized in Tables 2 and 3.

Table 2. Donor numbers (DN) of selected solvents and anions

Donor	DN	Donor	DN
nitromethane	2.7[a]	pyridine	33.1[a]
acetonitrile	14.1[a]	NCS^-	34[b]
I^-	15[b]	HMPTA[c]	38.8[a]
acetone	17.0[a]	F^-, CN^-	39[b]
water	18[a]	$[Co(CN)_6]^{3-}$	40[d]
methanol	19[a]	N_3^-	41[b]
Br^-	20[b]	$[Ru(CN)_6]^{4-}$	41[d]
dimethylformamide	26.6[a]	$[W(CN)_8]^{4-}$	43[d]
Cl^-	27[b]	$[Mo(CN)_8]^{4-}$	47[d]
diemthylsulfoxide	29.8[a]	piperidine	51[a]

[a] see [32], p. 141, [b] see [33], p. 62, [c] hexamethylphosphoric triamide, [d] unpublished results

According to Tables 2 and 3, the methyl viologen dication (MV^{2+}) is a weaker acceptor than water, whereas $[Co(NH_3)_6]^{3+}$ exhibits a much stronger σ-acceptor strength. On the other hand, cyanometallates are stronger electron-pair donors than water. This explains, at least qualitatively, why the experimental values of K_{IP} for the ion pair $MV^{2+}/[Fe(CN)_6]^{4-}$ coincide with the values calculated by using Eq. (3) [36] whereas the experimental value of the ion pair of $[Co(NH_3)_6]^{3+}$ with the same counterion exceeds the calculated one by a factor of ten [28]. The importance of electron-pair donor-acceptor interactions becomes particularly evident with respect to optically active ions. Thus, the ion-pair formation constants estimated for ion pairs of Λ-$[Co(en)_3]^{3+}$ and Δ-$[Co(en)_3]^{3+}$ respectively, with $[Sb_2(d\text{-tartrate})_2]^{2-}$ differ by a factor of 2 [37]. In the case of *R*,*R*-tartrate^{2-}, the ion-pair formation constants for the two ion pairs only differ by a factor of 1.3 [20].

Stereoselectivity is observed during the oxidation of cobalt(II) ammine complexes by $[Co(III)Y]^-$ (Y = EDTA, PDTA, CDTA) [38–41]. In all cases, the formation of the product originates from the ion pair with the strongest interaction between

Table 3. Acceptor numbers (AN) of selected solvents and cations

Acceptor	AN	Acceptor	AN
dimethylformamide	16.0[a]	Na^+	38[b]
dichloromethane	20.4[a]	$[Co(sep)]^{3+}$	40[d]
K^+	26.5[b]	methanol	41.3[a]
$[Co(en)_2Cl_2]^+$	29[d]	water	54.8[a]
MV^{2+}	29[d]	$[Co(en)_3]^{3+}$	59[d]
Ph_2I^+	30[c]	$C_7H_7^+$	66[d]
NH_4^+	31.7[b]	$[Co(NH_3)_6]^{3+}$	68[d]
i-propanol	33.5[a]	formic acid	83.6[a]
ethanol	37.1[a]	CF_3COOH	105.3[a]

[a] see [32], p. 141, [b] see [34], [c] see [35], [d] unpublished results

the planes of the coordination polyhedra is observed. The weaker the coordination of the solvent at the enantiomeric educt complex, the stronger is the domination of one enantiomer in the product mixture [38]. Different rates of formation of various enantiomeric cobalt(III) ammine complexes are explained by the following effects:

- increased formation constant of one distinguished stereoisomeric precursor pair (thermodynamic effect),
- differing orbital overlapping due to differences in the distances of reacting ions,
- differences in reorganization energies for the electron transfer in stereoisomeric ion pairs (kinetic effect).

The concerted action of these effects may favour the stereoselective formation of one product. However, these effects may also act in a converse manner [42].

It has been pointed out by Haim [28] that there is evidence for the existence of isomeric ion pairs where electron transfer can be much faster in the thermodynamically unstable ion pair than in the thermodynamically stable one. Quenching experiments [43] and IR studies [44] also provide evidence for the existence of isomeric ion pairs. The coexistence of various isomeric ion pairs between $[Fe(phen)_3]^{2+}$ and arene sulfonates has been demonstrated by means of NMR measurements [45].

Although the use of optically active ions offers an interesting insight into mechanistic details of the photolysis, only one paper dealing with this subject has been published so far [46].

2.4 Dynamics of Formation and Decay of Ion Pairs

Based on the electrostatic model (see Sect. 2.2.) the rate of the diffusion-controlled formation (k_d) [47, 48] and decay (k_{-d}) [27] of ion pairs may be described by the Debye-Smoluchowski (4) and Eigen (5) equations

$$k_d = \frac{2Nk_BT}{3000\eta}\left(2 + \frac{r_+}{r_-} + \frac{r_-}{r_+}\right)\left(\frac{-w}{\exp(-w) - 1}\right) \tag{4}$$

$$k_{-d} = \frac{k_BT}{2\pi\eta}\left(\frac{1}{(r_+ + r_-)(r_+ \cdot r_-)}\right)\left(\frac{-w}{1 - \exp(w)}\right) \tag{5}$$

where

$$w = \frac{z_+ \cdot z_- e^2}{(r_+ + r_-)D_s k_B T}$$

η — solvent viscosity
k_B — Boltzmann constant
r_-, r_+ — radii of anion and cation, respectively

In Eqs. (4) and (5), the work term w is a function of the ionic strength [49]. The validity of these equations has been confirmed for the ion pair $[Co(NH_3)_6]^{3+}$; SO_4^{2-} by means of ultrasound measurements [21]. The experimental data estimated to $k_d = 2.4 \times 10^{11}\ M^{-1}\ s^{-1}$ and $k_{-d} = 2 \times 10^8\ s^{-1}$ are in reasonable agreement with those calculated, $10^{11}\ M^{-1}\ s^{-1}$ and $3.3 \times 10^8\ s^{-1}$, respectively.

As long as no specific steric requirements are to be met, the dynamics and energetics of the formation of contact ion pairs do not differ from those of encounter complexes [10]. However, marked differences are observed if the energetically favoured conformation for the formation of contact ion pairs is reached only slowly. In this case, the ion pair formation is no longer considered to be diffusion-controlled. This may be illustrated by the ion pair $Ru(bpy)_3]^{2+}$; $[Mn(OH)PW_{11}O_{39}]^{6-}$ [43] for which partial entering of the cation into cavities of the anion is discussed. This interaction leads to different rate constants for dynamic and static quenching processes of the excited ruthenium complex.

Different reactivities of ion pairs between dyestuff cations and arylamino-methane sulfonates and the corresponding encounter complexes due to different positions of the anions with respect to the excited dyestuff cation have also been observed [50, 51]. Unusually slow formation of ion pairs ($k = 3.1 \times 10^{-3}\ M^{-1}\,s^{-1}$), has been found for the ion pair MV^{2+}; $TFBP^-$ [52] (see also Sect. 5). This effect is interpreted in terms of steric hindrance of the sandwich-like overlap of phenyl rings of the anion with the π-system of the cation [53].

3 Spectroscopy of Ion Pairs

3.1 The Ion-Pair Charge-Transfer Phenomenon

In addition to electrostatic attractive forces and electron-pair donor-acceptor interactions, ion pairing, depending on the relative energies of the highest occupied and lowest unoccupied orbital, respectively, may lead to electronic interactions that result in the formation of novel electronic states within the ion pair. Due to the weak electronic coupling, the energies of these novel electronic states are mainly determined by the difference in redox potentials of the participating ions [7].

If these differences are small, additional electronic transitions which correspond to an electron transfer from the donor ion to the acceptor ion may occur. These electronic transitions, which have been observed for the first time by Linhard [54] in the system $[Co(NH_3)_6]^{3+}$; I^- are named ion-pair charge-transfer (IPCT) transitions.

Examples for ion pairs of coordination compounds showing IPCT absorption bands are summarized in Table 4. Even in such cases where no IPCT bands can be seen (absorption in the short-wavelength UV region, superposition with more intense absorption bands of the individual ions), IPCT interactions have an influence on the photophysical, photochemical and other spectroscopic properties of ion pairs [137] (see Tables 1 and 5).

Table 4. Spectroscopic data of ion pairs with IPCT behaviour

Cation	Anion	Solvent	ΔG_{CT} [a]	ε_{max} [b]	$\Delta\bar{\nu}_{1/2}$ [a]	K_{IP} [c]	Ref.
$[Co(NH_3)_6]^{3+}$	$[Ru(CN)_6]^{4-}$	H_2O	29.2	243			[55]
$[Co(NH_3)_6]^{3+}$	$[Ru(CN)_6]^{4-}$	DMSO	28.8	580	10.87		[56]
$[Co(NH_3)_6]^{3+}$	$[Mo(CN)_8]^{4-}$	H_2O	24.7				[57]
$[Co(NH_3)_6]^{3+}$	$[Fe(CN)_6]^{4-}$	H_2O	22.7	300			[28]
$[Co(NH_3)_6]^{3+}$	I^-	H_2O, I = 0.06 M	37.3	2300	5.0	3.6	[58]
$[Co(NH_3)_6]^{3+}$	I^-	CH_3CN//	28.0				[59]
$[Co(NH_3)_6]^{3+}$	$S_2O_3^{2-}$	H_2O, I = 9M	37.5	1200			[60]
$[Co(NH_3)_6]^{3+}$	$SeCN^-$	H_2O	35.1				[57]
$[Co(NH_3)_6]^{3+}$	N_3^-	H_2O	35.8				[57]
$[Co(NH_3)_6]^{3+}$	$C_2O_4^{2-}$	H_2O	41.2				[57]
$[Co(NH_3)_6]^{3+}$	BPh_4^-	CH_2Cl_2/CH_3OH	31.3				[61]
$[Co(NH_3)_5py]^{3+}$//	$[Fe(CN)_6]^{4-}$	H_2O, pH 5, I = 0.1M	22.5	160		2400	[62]
$[Co(en)_3]^{3+}$	$[Ru(CN)_6]^{4-}$	H_2O	28.7				[55]
$[Co(en)_3]^{3+}$	$[Ru(CN)_6]^{4-}$	DMSO	26.4				[63]
$[Co(en)_3]^{3+}$	$[Fe(CN)_6]^{4-}$	H_2O, I = 2M	22.2	180	4.0	2.5	[64]
$[Co(en)_3]^{3+}$//	I^-	H_2O	35.0				[65]
$[Co(en)_3]^{3+}$	I^-	H_2O//	34.5				[57, 63]
$[Co(en)_3]^{3+}$	$SeCN^-$	H_2O	34.7				[57]
$[Co(en)_3]^{3+}$	$S_2O_3^{2-}$	H_2O	33.9				[57]
$[Co(en)_3]^{3+}$	N_3^-	H_2O	34.6				[57]
$[Co(en)_3]^{3+}$	$C_2O_4^{2-}$	H_2O	39.7				[57]
$[Co(en)_3]^{3+}$//	htc^-	H_2O	>30				[66]
$[Co(en)_3]^{3+}$	BPh_4^-	CH_3OH/CH_2Cl_2	29.4	520			[67]
$[Co(sep)]^{3+}$//	$[Ru(CN)_6]^{4-}$	H_2O	26.9				[55]
$[Co(sep)]^{3+}$	$[Fe(CN)_6]^{4-}$	H_2O	23.3	106			[28]
$[Co(sep)]^{3+}$	I^-	H_2O	34.6				[68, 69]
$[Co(sep)]^{3+}$	$SeCN^-$	H_2O	32.4				[57]
$[Co(sep)]^{3+}$	$S_2O_3^{2-}$	H_2O	32.4				[57]
$[Co(sep)]^{3+}$	NO_2^-	H_2O	31.4				[57]
$[Co(sep)]^{3+}$	N_3^-	H_2O	32.0				[57]
$[Co(sep)]^{3+}$	SCN^-	H_2O	35.2				[68, 69]

$[Co(sep)]^{3+}$	Br^-	H_2O	36.8				[68, 69]
$[Co(sep)]^{3+}$	$C_2O_4^{2-//}$	H_2O	36.4				[69, 70]
$[Co(sep)]^{3+}$	OCN^-	H_2O	36.6				[57]
$[Co(sep)]^{3+}$	Cl^-	H_2O	38.0				[68]
$[Co(sep)]^{3+}$	BPh_4^-	THF	31.3	900			[71]
$[Co(tacn)_2]^{3+}$	$[Ru(CN)_6]^{4-}$	H_2O	27.2				[57]
$[Co(tacn)_2]^{3+}$	I^-	H_2O	32.4				[57]
$[Co(tacn)_2]^{3+}$	$SeCN^-$	H_2O	32.6				[57]
$[Co(tacn)_2]^{3+}$	$S_2O_3^{2-}$	H_2O	32.8				[57]
$[Co(tacn)_2]^{3+}$	N_3^-	H_2O	31.3				[57]
$[Co(tacn)_2]^{3+}$	NO_2^-	H_2O	31.9				[57]
$[Co(tacn)_2]^{3+}$	SCN^-	H_2O	35.1				[57]
$[Co(tacn)_2]^{3+}$	$C_2O_4^{2-}$	H_2O	36.6				[57]
$[Co(tacn)_2]^{3+}$	OCN^-	H_2O	36.3				[57]
$[Co(tacn)_2]^{3+}$	Br^-	H_2O	38.6				[57]
$[Co(tacn)_2]^{3+}$	CH_3COO^-	H_2O	38.3				[57]
$[Co(bzo_3tacn)_2]^{3+}$	I^-	H_2O, I = 0.5M	25.3			2.5	[72]
$[Co(bzo_3tacn)_2]^{3+}$	NO_2^-	H_2O	21.8			1.7	[72]
$[Co(bzo_3tacn)_2]^{3+}$	SCN^-	H_2O	27.7			3.3	[72]
$[Co(phen)_3]^{3+}$	I^-	H_2O	27.4			40	[72]
$[Co(phen)_3]^{3+}$	NO_2^-	H_2O	22.1			10	[72]
$[Co(phen)_3]^{3+}$	SCN^-	H_2O	27.7			37	[72]
$[Co(phen)_3]^{3+}$	$C_2O_4^{2-}$	H_2O	32.1				[57]
$[Co(bpy)_3]^{3+}$	I^-	H_2O	26.8				[57]
$[Co(bpy)_3]^{3+}$	$C_2O_4^{2-}$	H_2O	31.8				[57]
$[Co_2(fulvalene)_2]^{2+}$	BPh_4^-	CH_3CN	24			26.3	[73]
$[Cp_2Co]^+$	$[Co(CO)_4]^-$	THF//	19.2				[74]
$[Cp_2Co]^{+//}$	$[Mn(CO)_5]^-$	solid	13.5				[75]
$[Co(CO)_3(PBu_3)_2]^+$	$[Co(CO)_4]^-$	THF//	25.0				[76]
$[Co(CO)_3(PPh_3)_2]^+$	$[Co(CO)_4]^-$	acetone	25.9	2500			[77]
$[Ru(NH_3)_6]^{3+}$	$[Ru(CN)_6]^{4-}$	solid	18.2				[78]
$[Ru(NH_3)_6]^{3+}$	$[Fe(CN)_6]^{4-}$	solid	13.5				[78]
$[Ru(NH_3)_6]^{3+}$	$[Fe(CN)_5CO]^{3-//}$	solid	22.2				[79]
$[Ru(NH_3)_6]^{3+}$	$[Rh(CN)_6]^{3-}$	H_2O	33.7	29			[80]
$[Ru(NH_3)_6]^{3+}$	I^-	H_2, I = 0.5M	24.9	244	6.4	9.8	[81]

Table 4. (continued)

Cation	Anion	Solvent	ΔG_{CT} [a]	ε_{max} [b]	$\Delta\bar{\nu}_{1/2}$ [a]	K_{IP} [c]	Ref.
$[Ru(NH_3)_6]^{3+}$	Br^-	H_2O	31.0	222	5.6	11.1	[81]
$[Ru(NH_3)_6]^{3+}$	Cl^-	H_2O	34.0	247		15	[81]
$[Ru(NH_3)_6]^{3+}$	SCN^-	H_2O	26.8				[72]
$[Ru(NH_3)_6]^{3+}$	SCN^-	H_2O, I = 0.01M	30.8				[82]
$[Ru(NH_3)_6]^{3+}$	$S_2O_3^{2-}$	H_2O	24.8				[82]
$[Ru(NH_3)_6]^{3+}$	CN^-	H_2O	24.6	174			[80]
$[Ru(NH_3)_5py]^{3+}$	$[Ru(CN)_6]^{4-}$	solid	14.5				[78]
$[Ru(NH_3)_5py]^{3+}$ //	$[Ru(CN)_6]^{4-}$	H_2O, pH 5, I = 0.1M	15.6	38	7.1	2900	[83]
$[Ru(NH_3)_5py]^{3+}$ //	$[Os(CN)_6]^{4-}$	H_2O	15.3	40	9.0	2700	[83]
$[Ru(NH_3)_5py]^{3+}$	$[Fe(CN)_6]^{4-}$	solid	10.2				[78]
$[Ru(NH_3)_5py]^{3+}$ //	$[Fe(CN)_6]^{4-}$	H_2O, pH 5, I = 0.1M	11.0	33	6.3	2500	[83]
$[Ru(NH_3)_5py]^{3+}$	I^-	H_2O, pH 4, I = 0.1M	24.4				[84]
$[Ru(NH_3)_5py]^{3+}$	Br^-	H_2O	29.6				[84]
$[Ru(NH_3)_5py]^{3+}$	SCN^-	H_2O	25.0				[84]
$[Ru(NH_3)_5py]^{3+}$	Cl^-	H_2O	32.1				[84]
$[Ru(NH_3)_5py]^{3+}$	$C_2O_4^{2-}$	H_2O	24.5				[84]
$[Ru(NH_3)_5CH_3CN]^{3+}$	Br^-	H_2O	30.0				[84]
$[Ru(NH_3)_5CH_3CN]^{3+}$	Cl^-	H_2O	31.5				[84]
$[Ru(en)_3]^{3+}$	I^-	H_2O	22.2				[85]
$[Ru(en)_3]^{3+}$	Br^-	H_2O	27.0				[85]
$[Ru(NH_3)_5Cl]^{2+}$	$[Ru(CN)_6]^{4-}$	H_2O	19.6	20	6.6	216	[86]
$[Pt(NH_3)_5Cl]^{3+}$	$[Ru(CN)_6]^{4-}$	H_2O	31.1	78			[87]
$[Pt(NH_3)_5Cl]^{3+}$	$[Os(CN)_6]^{4-}$	H_2O	28.4	140			[87]
$[Pt(NH_3)_5Cl]^{3+}$	$[Fe(CN)_6]^{4-}$	H_2O	23.9	88			[87]
$[Pt(NH_3)_5Cl]^{3+}$	$[Pt(CN)_4]^{2-}$	H_2O	33.9				[87]
$[Os(NH_3)_5Cl]^{2+}$	$[Ru(CN)_6]^{4-}$	H_2O	26.9		7.2		[55]
$[Os(NH_3)_5Cl]^{2+}$	$[Os(CN)_6]^{4-}$	H_2O	25.8		7.5		[55]
$[Os(NH_3)_5Cl]^{2+}$	$[Fe(CN)_6]^{4-}$	H_2O	22.8		5.6		[55]
$[Rh(bpy)_3]^{3+}$	$[Ru(CN)_6]^{4-}$	H_2O	26.4	110			[88]
$[Rh(bpy)_3]^{3+}$	$[Os(CN)_6]^{4-}$	H_2O	25.0	155			[88]
$[Rh(bpy)_3]^{3+}$	$[Fe(CN)_6]^{4-}$	H_2O	20.8	61		3800	[88]

$[Eu(2.2.1.)]^{3+}$	$[Ru(CN)_6]^{4-}$	H_2O, I = 1M	23.0	120	6.5	400	[89, 90]
$[Eu(2.2.1.)]^{3+}$	$[Os(CN)_6]^{4-}$	H_2O	22.2	110	7.8	800	[89, 90]
$[Eu(2.2.1.)]^{3+}$	$[Fe(CN)_6]^{4-}$	H_2O	18.9	110	6.1	300	[89, 90]
$[Ni(tim)]^{2+}$	$[Pt(mnt)_2]^{2-}$	solid	12.1				[91]
$[Ni(tim)]^{2+}$	$[Pd(mnt)_2]^{2-}$	solid	12.0				[91]
$[Ni(tim)]^{2+}$	$[Ni(mnt)_2]^{2-}$	solid	11.9				[91]
$[Cr(phen)_3]^{3+//}$	$I^{-//}$	solid	25				[92]
$[Fe(bpy)_3]^{2+//}$	$TCNE^{-\cdot}$	solid	26.0				[93]
$[Fe(C_6Me_6)_2]^{2+//}$	iso-$[C_4(CN)_6]^{2-//}$	solid	20.0				[94]
$TROP^+$	$[Ru(CN)_6]^{4-}$	$H_2O^{//}$	22.0				[63]
$TROP^+$	$[Mo(CN)_8]^{4-}$	$H_2O^{//}$	18.6				[63]
$TROP^+$	$[Mn(CN)_5NO]^{3-}$	H_2O	18.4				[63]
$TROP^+$	$[W(CN)_8]^{4-}$	$H_2O^{//}$	16.3				[63]
$TROP^+$	$[Fe(CN)_6]^{4-}$	$H_2O^{//}$	18.0				[63]
$TROP^+$	$I^{-//}$	$CH_2Cl_2^{//}$	17.4	1880			[95–98]
$TROP^+$	Br^-	CH_2Cl_2	24.4	1380			[96]
$TROP^+$	Cl^-	CH_2Cl_2	33.6	1820			[96]
$TROP^+$	SCN^-	CH_2Cl_2	20.7				[99]
$TROP^+$	$(CO_2CH_3)_5C_5^-$	$CH_2Cl_2^{//}$	21.3				[100, 101]
Ph_2I^+	$[Ru(CN)_6]^{4-}$	H_2O	32				[57]
Ph_2I^+	$[Ru(CN)_6]^{4-}$	$CH_3OH^{//}$	29.4	1000	8.0		[102]
Ph_2I^+	$[Fe(CN)_5DMSO]^{3-}$	CH_3OH	30.0	1400	7.2		[102]
Ph_2I^+	$[Mo(CN)_8]^{4-}$	H_2O	28.5				[57]
Ph_2I^+	$[Mo(CN)_8]^{4-}$	$CH_3OH^{//}$	27.0	1050	8.6		[102]
Ph_2I^+	$[Mn(CN)_5NO]^{3-}$	CH_3OH	25.7	980	6.6		[102]
Ph_2I^+	$[W(CN)_8]^{4-}$	H_2O	27.4				[57]
Ph_2I^+	$[W(CN)_8]^{4-}$	CH_3OH	25.0	1350	8.8		[102]
Ph_2I^+	$[Fe(CN)_6]^{4-}$	H_2O	27.1				[57]
Ph_2I^+	$[Fe(CN)_6]^{4-}$	CH_3OH	24.6		7.6		[102]
Ph_2I^+	I^-	$CH_3CN^{//}$	30.9	1200			[59, 103, 104]
$C_6H_5{-}N_2^+$	I^-	H_2O	28.9	380		35	[104]
p$-CH_3O-C_6H_4-N_2^+$	$[Ru(CN)_6]^{4-}$	H_2O	26.7				[105]
p$-CH_3O-C_6H_4-N_2^+$	$[Mo(CN)_8]^{4-}$	H_2O, pH 7, I = 0.1M	25.0				[57]
KOS^+	$[Ru(CN)_6]^{4-}$	$CH_3OH^{//}$	22.7				[63]
KOS^+	$[Mo(CN)_8]^{4-}$	$H_2O^{//}$	22.5				[63]

Table 4. (continued)

Cation	Anion	Solvent	ΔG_{CT} [a]	ε_{max} [b]	$\Delta\bar{v}_{1/2}$ [a]	K_{IP}	Ref.
KOS^+	$[Mn(CN)_5NO]^{3-}$	H_2O	23.3				[63]
KOS^+	$[W(CN)_8]^{4-}$	H_2O//	20.4				[63]
KOS^+	$[Fe(CN_6]^{4-}$	H_2O//	21.5				[63]
KOS^+//	$[Ni(mnt)_2]^{2-}$//	solid	14.8				[267]
KOS^+	$[Cu(mnt)_2]^{2-}$	solid	14.0				[267]
KOS^+//	$[Zn(mnt)_2]^{2-}$	solid	21.5				[267]
KOS^+//	$[Co(mnt)_2]^{2-}$	solid	14.8				[267]
KOS^+//	I^-//	$CHCl_3$//	22.3	1230			[106–109]
MV^{2+}	$[Ru(CN)_6]^{4-}$	H_2O	24.0				[63]
MV^{2+}	$[Fe(CN)_5DMSO]^{3-}$//	H_2O	22.2				[36]
MV^{2+}	$[Mo(CN)_8]^{4-}$	H_2O	19.8				[63]
MV^{2+}	$[Mn(CN)_5NO]^{3-}$	H_2O	19.5				[63]
MV^{2+}	$[W(CN)_8]^{4-}$	H_2O	17.4				[63]
MV^{2+}//	$[Fe(CN)_6]^{4-}$	H_2O, $I = 0.1M$	18.9	180		36	[110, 111]
MV^{2+}	$[Fe(CN)_6]^{4-}$	H_2O, pH 5, $I = 0.1M$	18.9	50	6.8	220	[112]
MV^{2+}//	$[Zn(mnt)_2]^{2-}$//	$DMSO/CHCl_3$//	19.1	404		683	[113]
MV^{2+}	$[Zn(mnt)_2]^{2-}$	CH_3CN	20.0				[114]
MV^{2+}//	$[Pt(mnt)_2]^{2-}$	DMF	11.8	30		240	[115]
MV^{2+}	$[PbI_4]^{2-}$//	solid	19.0				[116]
MV^{2+}	I^-//	H_2O, pH 7	28.6				[117]
MV^{2+}	BPh_4^-	CH_3CN	31.1	151			[118]
MV^{2+}	$C_2O_4^{2-}$//	H_2O//	>25				[119–122]
MV^{2+}	H_2EDTA^{2-}	H_2O, pH 8	>28			13	[123]
MV^{2+}	$(EtO)_2PS_2^{2-}$	CH_3CN	25				[124]
MV^{2+}	$TFPB^-$	DME	20.8	110		2110	[52, 125]
pyH^+//	$[V(CO)_6]^-$	solid	red				[126]
pyH^+//	$[Co(CO)_4]^-$	butan-2-one	24.1				[19]
$DMBPY^{2+}$	$[Co(CO)_4]^-$	CH_2Cl_2	21.3	560		120	[127]
TPP^+//	I^-//	CH_2Cl_2//	18.0	400			[128, 129]
$Lucigenin^{2+}$	NO_3^-//	C_2H_5OH	>20				[130]
Bu_4N^+	I^-	CCl_4	34.5	17500			[131]

Ph_3S^+	I^-	CH_3OH	>33			[132]
$(4-CN-C_6H_4)(CH_3)_3N^+$	I^-	$CHCl_3$	34.5	870		[133]
CV^+	I^- //	H_2O	28.3	3100	30	[134]
$[Co(EDTA)]^-$	I^-	H_2O	34.6			[135]
$[Fe(CN)_6]^{3-}$	$[Fe(CN)_6]^{4-}$	H_2O	12.9			[136]

[a] in $10^3\ cm^{-1}$; [b] in $L\ mol^{-1}\ cm^{-1}$; [c] in $L\ mol^{-1}$
// data of similar systems are included in the same reference

Table 5. Comparison of calculated and measured cage-escape efficiencies for ion pairs A^+; B^- in aqueous solution after IPCT excitation

A^+	B^-	$r_{A\cdot}$ [pm][a]	$r_{B\cdot}$ [pm][b]	k_{-d} [$10^9\ s^{-1}$][c]	$\eta_{-d,\ calc.}$[d]	$\eta_{-d,\ exp.}$[e]	Ref.
$[Co(NH_3)_6]^{3+}$	I^-	370	133	26	0.21	0.19	[162]
$[Co(NH_3)_6]^{3+}$	Cl^-	370	99	38	0.28	0.30	[163]
$[Rh(bpy)_3]^{3+}$	$[Fe(CN)_6]^{4-}$	700	430	0.17	0.0017	0.0024	[88]
$4-CH_3O-C_6H_4-N_2^+$	$[Ru(CN)_6]^{4-}$	350[f]	450	5.2	0.049	0.02	[105]

[a] from Ref. [164], [b] from Ref. [138], [c] calculated according to Eq. (5), [d] $\eta_{-d} = k_{-d}/(k_{back} + k_{-d})$ and $k_{back} = 10^{11}\ s^{-1}$, [e] corresponds to Φ_{IPCT}, [f] estimated

3.2 The Position of IPCT Absorption Bands

A more detailed description of the IPCT phenomenon requires the knowledge of the energetic situation. A convenient descriptive model has been proposed by Cannon [137, 138]. This model is based upon three assumptions:

First it is assumed that the charge delocalization, which occurs during the formation of the precursor ion pair $(A^+; B^-)_s$ from the solvated individual ions A_s^+ and B_s^- and which is responsible for the IPCT absorption, is negligible. The same holds true for the successor pair $(A; B)_s$.

Secondly, the Franck-Condon excited state of the successor pair $(A; B)_{s, FC}$ is considered as a thermodynamic state, i.e. it is assumed that the deviation from the Boltzmann distribution, due to the circumstance that the Franck-Condon excited state is not in an equilibrium with its surrounding, can be neglected.

At third, the model is based upon the assumption that there is no change in entropy during light absorption, i.e. $\Delta G_{CT} = E_{CT} = h\nu_{CT}$.

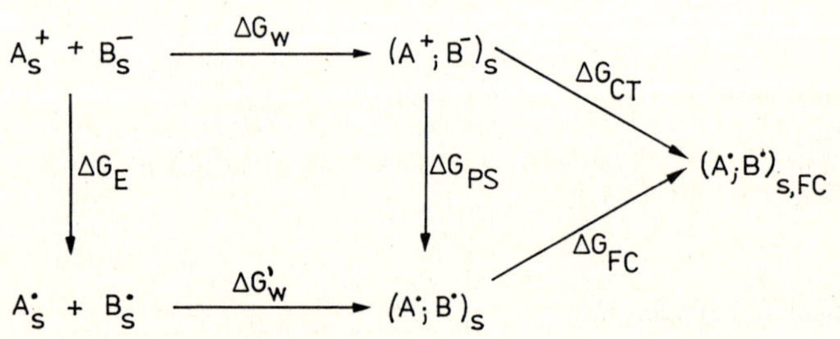

From Scheme 1, one may deduce Eq. (6):

$$\Delta G_{CT} = \Delta G_E + \Delta G_{W'} - \Delta G_{W'} + \Delta G_{FC} = \Delta G_{PS} + \Delta G_{FC} \tag{6}$$

The redox asymmetry ΔG_E is calculated from the standard redox potentials of the participating ions according to Eq. (7)

$$\Delta G_E = F[E^0(D/D^-) - E^0(A^+/A)] \tag{7}$$

(F = Faraday constant)

The free enthalpies of association ΔG_W and $\Delta G_{W'}$, respectively, can be estimated from the ionic radii (see Sect. 1). For the calculation of the contribution of Franck-Condon energy, which may be separated into an inner-sphere and an outer contribution, various approaches have been developed [10, 49, 137–139].

Although the Cannon model allows, in principle, the prediction of the energy ΔG_{CT} of the absorption maximum of the IPCT band, the low accuracy of the required electrochemical, vibrational spectroscopy and X-ray data for the oxidized and reduced ions leads to satisfactory results only in few cases [63].

An increment system based on the known spectroscopic data of IPCT systems has been proposed [63] in order to predict the energies of IPCT bands.

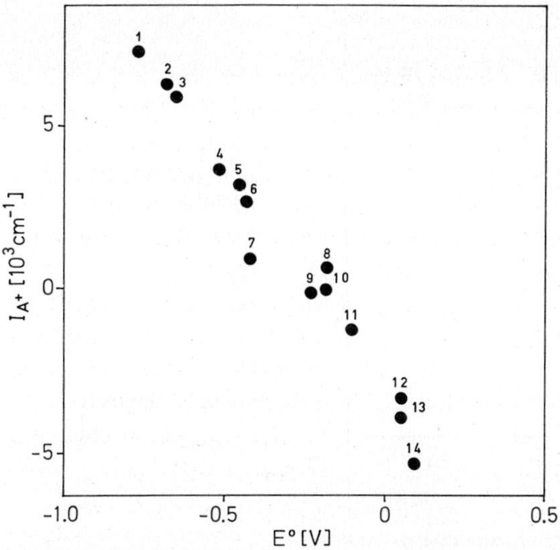

Fig. 1. Increments of cations A^+ vs their reduction potentials
(1) $[Co(NH_3)_6]^{3+}$; *(2)* $[Co(en)_3]^{3+}$; *(3)* $[Co(sep)]^{3+}$; *(4)* $[Co(tacn)_2]^{3+}$; *(5)* MPZ^+; *(6)* $[Rh(bpy)_3]^{3+}$; *(7)* MV^{2+}; *(8)* $[Eu(2.2.1.)]^{3+}$; *(9)* MQX^+; *(10)* $TROP^+$; *(11)* $[Co(bpy)_3]^{3+}$; *(12)* $[Co(bzo_3tacn)_2]^{3+}$; *(13)* $[Ru(NH_3)_6]^{3+}$; *(14)* $[Ru(NH_3)_5ImH]^{3+}$

The linearity between the oxidation and reduction potentials and increments of anions and cations, respectively, expected from Eqs. (6) and (7) has been observed for a large variety of structurally different anions and cations.

Using Eq. (7), it is possible to estimate spectroscopically not accessible increments from the redox potentials of corresponding ions [63].

Using the increment system it is also possible to predict the energies of absorption bands of "inverse" IPCT systems, i.e. electronic transitions from a donor cation to an acceptor anion. Interestingly, this type of transition has not yet been observed in solution. It is, however, assumed that in solid silver(I) permanganate additional absorption bands may be assigned to an electronic transition from the silver cation to the permanganate anion [140].

3.3 The Shape of IPCT Absorption Bands

There is a dependence of the extinction coefficient of the IPCT absorption band on the frequency ν of the absorbed light which may be described by Eq. (8) given by Ulstrup [111]

$$\varepsilon(\nu) = \frac{2\pi}{3 \cdot 2.3\,c} (M_{eg})^2 \frac{h\nu}{k_B T h} K(\nu) \quad (8)$$

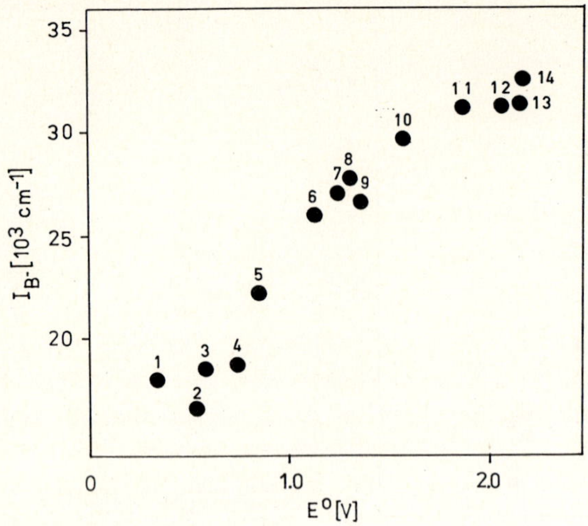

Fig. 2. Increments of anions B^- vs. their oxidation potentials
(1) $[Fe(CN)_6]^{4-}$; *(2)* $[W(CN)_8]^{4-}$; *(3)* $[Mn(CN)_5NO]^{3-}$; *(4)* $[Mo(CN)_8]^{4-}$; *(5)* $[Ru(CN_6]^{4-}$; *(6)* NO_2^-; *(7)* $SeCN^-$; *(8)* I^-; *(9)* N_3^-; *(10)* SCN^-; *(11)* Br^-; *(12)* $C_2O_4^{2-}$; *(13)* OCN^-; *(14)* Cl^-

Here, M_{eg} is the dipole matrix element of the electronic transition that determines the intensity of the absorption band, $K(\nu)$ is a function that describes the band shape. Within the range of one half-width the band shape function can be approximated by the Gauss function, Eq. (9) [111]. In equation (9), ν_{max} is the frequency of the absorption maximum of the IPCT band and $\Delta\nu_{1/2}$ is half-width of the absorption band

$$K(\nu) = \frac{2\sqrt{\pi k_B T}}{\Delta\nu_{1/2}} \exp\left[-\frac{(h\nu - h\nu_{max})^2}{(\Delta\nu_{1/2})^2}\right] \tag{9}$$

Usually, IPCT absorption bands are very broad ($\Delta\bar{\nu}_{1/2} = 4000$–$1000\ cm^{-1}$) and of medium intensity ($\varepsilon = 10$–$2000\ L\ mol^{-1}\ cm^{-1}$), see Table 4. At room temperature the half-width of the absorption band is linearly dependent on the square root of the temperature and the Franck-Condon energy ΔG_{FC} of the electron transfer, Eq. (10), [111, 141].

$$\Delta\nu_{1/2} = A\sqrt{\Delta G_{FC} k_B T} \tag{10}$$

Due to the "freezing" of solvent vibrations, a temperature-independent linewidth is predicted for the low-temperature region [111]. Experimentally, this could be confirmed for the ion pair $MV^{2+}/[Fe(CN)_6]^{4-}$ [111].

The factor A in Eq. (10) has been estimated by Hush [141] as $(16 \ln 2)^{1/2} = 3.33$, while more recent approaches [111] use a factor of 2. However, in most cases measured linewidths are significantly larger than those estimated by using Eq. (10) [55, 83, 87, 111, 112]. It is assumed [83] that both spin-orbital coupling in coordination compounds and the broad distribution of distances within the ion pairs due to weaker bonding forces as compared with covalent bonds, are the main reasons for the large linewidths observed for IPCT bands.

3.4 The Solvent-Dependence of IPCT Bands

Based on the Cannon model (see Sect. 3.1.), which describes the energetics of the IPCT phenomenon, the energy ΔG_{CT} of the IPCT absorption maximum may be separated into the contributions ΔG_{PS} and ΔG_{FC}. Here, ΔG_{PS} is dependent on both the redox potentials of the partners in the ion pair and the free energy of association. As outlined in Sect. 2, the free energy of association can be described by using the so-called continuum model. However, in some cases specific solvent-solute interactions have to be taken into account. Since the difference in free energy of association for the precursor and successor pair is, at least for polar solvents, relatively small, as compared with the total energy ΔG_{CT} of the IPCT, we may renounce a detailed discussion of the free energy of association.

With respect to charge distribution, ion pairs are asymmetric systems. Therefore, one may expect a different influence of the solvent on the redox potentials of donor and acceptor ions, respectively. Particularly for strong electron-pair donors and acceptors the oxidized and reduced form of the ion pair are differently stabilized by solvation. Thus, there is a linear relationship between the acceptor number of the solvent [142, 143] and the potential of the redox couple $[Fe(CN)_6]^{3-/4-}$. On the other hand, the potential of the redox couple $[Co(en)_3]^{3+/2+}$ and the donor number of solvents [144, 145] are correlated.

The reorganizational energy ΔG_{FC} of the electron transfer within the ion pair consists of two contributions: an inner contribution $\Delta G_{FC,in}$ corresponding to the Franck-Condon energy due to geometry changes of the inner coordination sphere, and an outer contribution $\Delta G_{FC,out}$ which is caused by the surrounding solvent

$$\Delta G_{FC} = \Delta G_{FC,in} + \Delta G_{FC,out} \tag{11}$$

In the first approximation, $\Delta G_{FC,in}$ can be regarded as solvent-independent.

With respect to the calculation of reorganizational energy caused by the solvent, Marcus [146] has proposed an approach which is based on the model of the dielectric continuum. Essentially, this model treats the influence of the change of the dipolar momentum as the result of electron transfer on the nuclear polarization of the solvent molecules. The outer fraction of the reorganizational energy, $\Delta G_{FC,out}$, can be calculated by using Eq. (12)

$$\Delta G_{FC,out} = (1/D_{op} - 1/D_s)\, Y \tag{12}$$

Here, $D_{op} = n^2$ (n — refractive index) and D_s are the optical and the static dielectric constants of the solvent, respectively. For the factor Y, various approximations have been introduced [138, 139, 147]. For ion pairs consisting of approximately spherical ions i.e. octahedral and tetrahedral ions, the factor Y is given by Eq. (13)

$$Y = e^2(1/2r_1 + 1/2r_2 - 1/d)\ a \tag{13}$$

In Eq. (13), r_1 and r_2 are the radii of the ions, d is the interionic distance, i.e. $d = r_1 + r_2$, and e is the electronic charge. The factor a, which may range between zero and unity, depends on the electronic interaction between donor and acceptor electronic states. For ion pairs, the assumption $a = 1$ appears to be reasonable.

The linear relationship between $(1/D_{op} - 1/D_s)$ and $\Delta G_{FC,out}$ has been demonstrated experimentally for a large variety of mixed-valence compounds [148, 149]. However, in recent studies performed by Hupp [150] the dependence of $\Delta G_{FC,out}$ has been questioned. An increase of external pressure should lead to a decrease of D_s by a factor of 2 to 10, while D_{op} should increase by less than 30%. However the observed bathochromic shift of the IPCT band was by orders of magnitude less than expected according to Eq. (12). Since, for polar solvents, D_s is usually very large and, hence, $1/D_s$ is very small, Eq. (12) can be used without any restrictions.

The solvent influence on the position of the IPCT band can be described by the empirical Eq. (14) derived very recently [59]:

$$\Delta G_{CT} = C_0 + C_1 AN + C_2 DN + C_3(1/D_{op} - 1/D_s) \tag{14}$$

In a series of ion pairs of iodide, thiocyanate and octacyanomolybdate(IV), the significance of the dependence of the energy of the IPCT absorption maxima on the acceptor number AN of the solvent could be demonstrated by a nonlinear regressional analysis. On the other hand, the dependence on the donor number DN was not significant in any of the cases studied, despite the fact that solvent donor number effects dramatically the reduction potential of the ion $[Co(en)_3]^{3+}$ [145, 149]. One of the reasons of the lack of statistical significance could be the fact that the width of variation of the solvent used was too narrow. On the other hand, a further uncertainty may arise from redox potentials obtained by electrochemical measurements, which refer to the reduction of low spin cobalt(III) amine complexes to high-spin cobalt(II) complexes, while light-induced reduction of cobalt(III) complexes leads to low-spin Co(II) in the primary step (see also Sect. 5).

A significant correlation between the absorption maximum and the parameter $(1/D_{op} - 1/D_s)$ has been found for the ion pair between 1-ethyl-4-cyanopyridinium cation and iodide. Here, the factor C_3 (see Eq. (14)), which corresponds to Y in Eq. (12), is in good agreement with the value calculated on the basis of the dipole in a sphere model [137].

When using mixtures of solvents, more complex solvent effects are observed. This is mainly due to the asymmetry of ion pairs with respect to electron-pair donor-acceptor interactions leading to preferred solvation of ions by only one component of the solvent mixture. Consequently, different effects of the solvent components on the redox potentials of the individual ions are observed. Moreover, the reorganization of the solvatational sphere after electron transfer not only requires a change in solvent nuclear polarization but also an exchange of solvent molecules with the surrounding bulk solution.

3.5 Further Changes in the Electronic Spectrum Induced by Ion Pairing

The formation of contact ion pairs leads to only weak electron delocalization. As a result, the absorption spectra of the individual ions within the ion pair, except for the appearance of an additional IPCT band, are not markedly changed with respect to the free ions [102].

Changes in the absorption spectrum are observed only if ion pairing leads to a perturbation of the electronic system of the partners in the ion pair. This may be the case if, e.g. the ions have an extended π-electron system. Thus, the absorption bands of $[Ni(mnt)_2]^{2-}$ in the ion pair with $[Ru(bpy)_3]^{2+}$ are markedly shifted as compared with the free ion [151], whereas no changes are observed for ion pairs with MV^{2+} [115].

An influence of the counterion on both the intensity and position of the absorption bands has also been documented for ion pairs with porphyrin complexes [22, 25, 152, 153], phthalocyanines [154] and rose bengale [155]. Ion pairs of the stilbene-

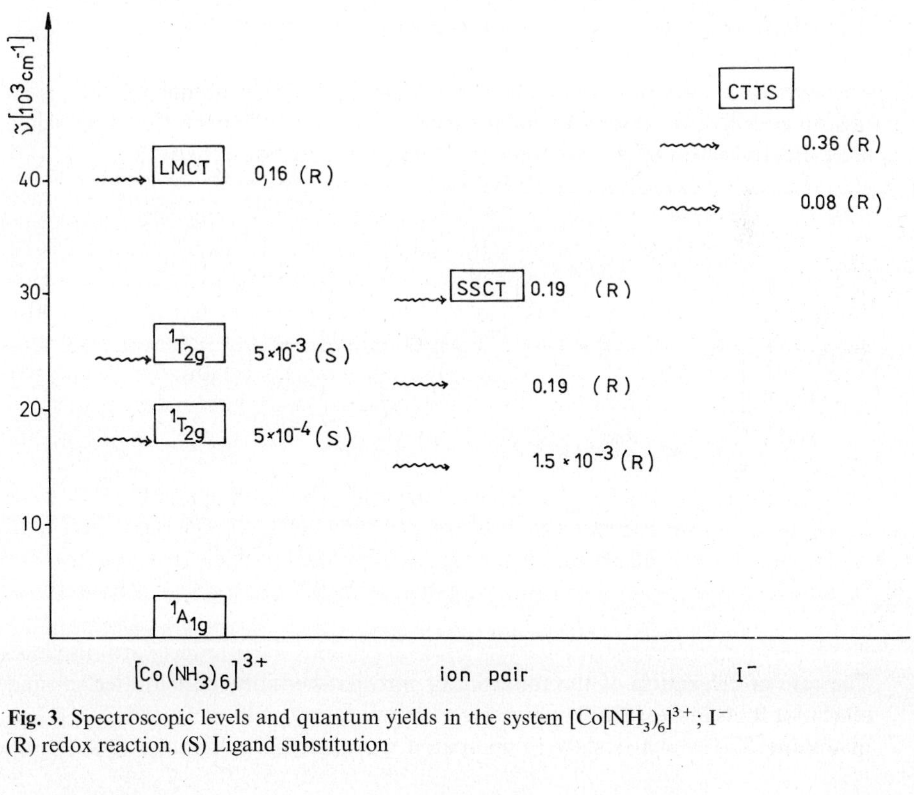

Fig. 3. Spectroscopic levels and quantum yields in the system $[Co[NH_3)_6]^{3+}$; I^- (R) redox reaction, (S) Ligand substitution

diazonium cation with tetra(3-trifluoromethylphenyl)borate exhibit a bathochromic shift of the absorption maximum of the diazonium ion accompanied by an unusual high tendency of ion pairing in aqueous solution [156].

Ion pairing may also lead to complete or partial thermal electron transfer which results in the formation of free radicals. Depending on the solvent and the temperature, both ion pairs, free radicals and free ions may be detected at the same time [98, 99].

4 Photochemistry of Ion Pairs

The photophysical and photochemical properties of the individual ions may be dramatically altered by ion pairing [7].

As an example, irradiation of $[Co(NH_3)_6]^{3+}$ and iodide ions is accompanied with photoreduction and photooxidation only upon excitation of charge-transfer bands in the short-wavelength UV region (see Fig. 3) whereas irradiation into the IPCT band leads to a photoredox process above 300 nm. Interestingly, photolysis at 500 nm, i.e. a region where absorption by the ion pair may be neglected, leads exclusively to redox products, even though the quantum yield is very low.

4.1 Photoreactions Induced by Irradiation into IPCT Absorption Bands

Photoinduced electron-transfer by irradiation into IPCT absorption bands of the ion pair $A^+ ; B^-$ leads to a radical pair $(A^\cdot ; B^\cdot)_{FC}$ in a Franck-Condon excited state, i.e. the same geometry as the precursor ion pair, see Scheme 2.

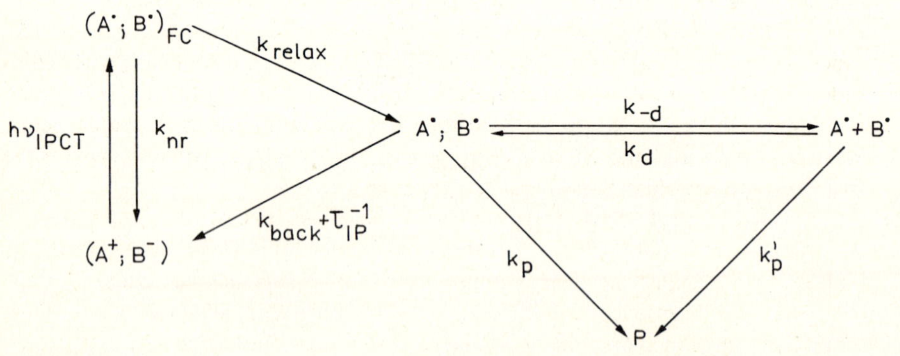

Scheme 2

The rate of relaxation of the radical pair into its thermally equilibrated ground state $(A^\cdot ; B^\cdot)$ is limited [207] by the dielectric relaxation of the solvent ($\tau_D = 8.7$ ps in water [157]) which is slow as compared with internal vibrations of the ions.

Radiationless deactivation into the ion-pair ground state via fast internal vibrations may occur as a competitive process. The requiremenf of such a radiationless deactivation is that there exists a set of internal vibrations which may accept the energy introduced into the system. This requirement is met if ΔG_{CT} is small and if there is a sufficient number of ligands in a not too rigid position. The efficiency $\eta_{relax} = k_{relax}/(k_{nr} + k_{relax})$ for the relaxation within the ion pair $[Ru(NH_3)_5py]^{3+}$; $[Fe(CN)_6]^{4-}$ ($\Delta G_{CT} = 10640\ cm^{-1}$) has been found by flash photolysis to be 0.06 [158]. However, to a certain extent this ion pair is exceptional since for the majority of ion pairs ΔG_{CT} is considerably larger than $10000\ cm^{-1}$ (see Table 4).

The successor pair (A·; B·) formed by relaxation may react in three different ways. The precursor ion pair (A^+; B^-) may be regenerated either by luminescence or by thermal back electron transfer whereas the solvated free radicals A· and B· may be formed by diffusive escape. In addition, there is the possibility of irreversible formation of products P by fast decay of one partner within the solvent cage.

Due to the low transition probabilities — the extinction coefficients for ion pair absorption bands are usually in the range of 10 to 2000 L $mol^{-1}\ cm^{-1}$ (see Table 4) — the luminescence lifetimes of relaxed excited IPCT states are relatively long. However, ion-pair luminescence is normally suppressed by fast diffusional processes and back electron transfer. In fact, the ion pair MV^{2+}; $TFPB^-$ is the only example for which ion-pair luminescence in solution has been reported at room temperature [159]. At low temperature, where radiationless deactivation processes are much slower, this phenomenon was observed earlier [160]

The cage escape efficiency, $\eta_{-d} = k_{-d}/(k_{-d} + k_p + k_{back} + 1/\tau_{IP})$, is normally determined by the ratio between the rate constant of the back electron transfer within the successor pair, k_{back}, and that of the diffusional escape of the successor pair, k_{-d}. An estimate of the rate of the back electron transfer may be made on the basis of the Marcus-Hush theory [28]. The maximum value is, however, limited by the dielectric relaxation of the solvent, i.e. $k_{back,\,max} = 1/\tau_D \approx 10^{11}\ s^{-1}$ [161]. The rate constant of the diffusional escape can be obtained from the Eigen Eq. (5).

Provided the back electron transfer is a strongly exergonic process that is determined by the rate of solvent relaxation and $\eta_{relax} = 1$, then it follows for ion pairs undergoing fast irreversible secondary reactions that the cage-escape efficiency, η_{-d}, is given by the rate constant of diffusion k_{-d}. For comparison, the experimental quantum yields and the efficiencies η_{-d} of a variety of ion pairs for which the above-mentioned condition is met are given in Table 5.

A reasonable agreement was found for a large number of different successor ion pairs strongly supporting the validity of the kinetic model. The smaller the radii and the smaller the charge numbers of the ions in the successor ion pair, the higher is the efficiency. However, η_{-d} does not exceed a value of 0.3 for relatively small ion pairs with low charges. An increase of efficiency is expected if the rate of back electron transfer is lowered by high Franck-Condon barriers. This also leads to a drastic hypsochromic shift of the IPCT absorption maximum.

Irreversible product formation may occur either by fast decomposition of one of the successor components inside the solvent cage or by secondary reactions after diffusional escape of the successor pair. The following processes may serve as competitive reactions leading to irreversible product formation:

Table 6. Photoreactions of ion pairs induced by light absorption by IPCT bands

Cation	Anion	Solvent	λ_{irr} [a]	Φ	Ref.
$[Co(NH_3)_6]^{3+}$	$[Ru(CN)_6]^{4-}$	DMSO	366	0.034	[56]
$[Co(en)_3]^{3+}$	$[Fe(CN)_6]^{4-}$	H_2O, 4M NaCl	436	0.46	[46]
$[Ru(NH_3)_5Cl]^{2+}$	$[Ru(CN)_6]^{4-}$	H_2O	546	0.002	[86]
$[Ru(NH_3)_5py]^{3+}$	$[Fe(CN)_6]^{4-}$	H_2O	1060	0.06[b]	[158]
$[Rh(bpy)_3]^{3+}$	$[Fe(CN)_6]^{4-}$	H_2O	546	0.0024	[88]
$[Co(CO)_3(PPh_3)_2]^+$	$[Co(CO)_4]^-$	acetone	405	0.012	[77]
$[Co(CO)_3(PBu_3)_2]^+$	$[Co(CO)_4]^-$	THF, PR_3	>300	[c]	[76]
$[Co(cp)_2]^+$	$[Co(CO)_4]^-$	THF, PR_3	520	0.3	[74]
$[Eu(2.2.1.)]^{3+}$	$[Fe(CN)_6]^{4-}$ //	H_2O	394	0	[89, 90]
$[Co(NH_3)_6]^{3+}$	I^-	H_2O, pH 4	370	0.19	[162, 167, 168]
$[Co(NH_3)_6]^{3+}$	I^-	H_2O	254	0.16	[163]
$[Co(NH_3)_6]^{3+}$	Cl^-	H_2O, 1M Cl^-	254	0.30	[163]
$[Co(NH_3)_6]^{3+}$	Cl^-	H_2O	254	0.21	[169]
$[Co(NH_3)_6]^{3+}$ //	BPh_4^-	CH_3OH/CH_2Cl_2	365	1.15	[61, 170]
$[Co(en)_3]^{3+}$	BPh_4^-	CH_3OH/CH_2Cl_2	365	0.21	[67]
$[Co(en)_3]^{3+}$	$C_2O_4H^-$	H_2O, pH 3	313	0.13	[171]
$[Co(en)_3]^{3+}$	H_2EDTA^{2-}	H_2O	254	0.24	[172]
$[Co(sep)]^{3+}$	I^-	H_2O, O_2 //	313	0.0016	[166]
$[Co(sep)]^{3+}$	I^-	H_2O, pH 7	313	0	[166]
$[Co(sep)]^{3+}$	Br^-	H_2O	313	0	[173]
$[Co(sep)]^{3+}$	$C_2O_4^{2-}$	H_2O, pH 5 //	313	0.29	[171]
$[Co(sep)]^+$	$C_2O_4H^-$	H_2O, pH 3	313	0.13	[171]
$[Co(sep)]^{3+}$	citrate^{3-}	H_2O · pH 8.5 //	313	0.11	[70]
$[Co(sep)]^{3+}$	BPh_4^-	THF	488	[c]	[71]
$[Co(EDTA)]^-$	I^-	H_2O	313	0.4	[135]
$[Co_2(fulvalene)_2]^{2+}$	BPh_4^-	CH_3CN	365	0.19	[73]
$[Ru(NH_3)_5py]^{3+}$	I^-	H_2O, pH 4	405	0.025	[84]
$[Ru(NH_3)_5py]^{3+}$	$C_2O_4H^-$	H_2O, pH 4	405	0.35	[84]
$DMBPY^{2+}$	$[Co(CO)_4]^-$	CH_2Cl_2 //, PPh_3	514.5	0.2	[127]
Ph_2I^+	$[Mo(CN)_8]^{4-}$	CH_3OH	475	0.51/5[d]	[102]
Ph_2I^+	I^-	H_2O	355	0.05	[104]

$4-CH_3O-C_6H_4-N_2^+$	$[Ru(CN)_6]^{4-}$	H_2O	405	0.02	[105]
$Quin^{+//}$	$[Co(CO)_4]^{-//}$	$THF^{//}$, PBu_3	550	0.44	[174]
CV^+	$I^{-//}$	$i-C_3H_7OH$	355	0	[134]
$Lucigenin^{2+}$	$NO_3^{-//}$	C_2H_5OH	>510	[c]	[130]
Ph_3S^+	I^-	$CHCl_3$	>320	0.35	[132]
$4-CH-C_6H_4N^+(CH_3)_3$	I^-	$CHCl_3$	254	0	[133]
KOS^+	I^-	benzene	>300	0.2[b]	[175]
MV^{2+}	BPh_4^-	CH_3CN	366	0.008	[118]
MV^{2+}	$TFPB^-$	solid	>365	[c]	[125, 176]
MV^{2+}	SCN^-	H_2O	337	0.3[b]	[177]
MV^{2+}	$Ph_2C(OH)COO^-$	CH_3CN	405	0.42	[120]
MV^{2+}	$Ph_2C(OH)COO^-$	H_2O	366	1.1	[119]
MV^{2+}	$C_2O_4^{2-}$	H_2O	340	0.24	[121]
MV^{2+}	H_2EDTA^{2-}	H_2O	366	0.062	[123, 178]
MV^{2+}	$PMAA^{n-}$	H_2O	355	0.051	[179]
MV^{2+}	$(CH_3O)_2PS_2^-$	CH_3CN	>400	0.003	[124]

[a] in nm, [b] cage-escape efficiency from flash-photolysis experiments, no permanent product formation, [c] product formation not quantitatively measured, [d] autocatalyzed reaction

// data of similar systems are included in the same reference

— reaction with a scavenger, e.g. solvent, one of the initial components or potential ligands,
— irreversible decomposition of one component of the successor pair,
— dimerization of components of the successor pair,
— secondary photoreactions of components of the successor pair at the same or different irradiation wavelength,
— irreversible change of spin states in the successor pair.

Although an irreversible change of the spin state cannot completely prevent the regeneration of the precursor ion pair by back electron transfer, it may cause a strong retardation of this process since it proceeds nonadiabatically. If there are products formed, which, as in the case of the ion pair $[Co(sep)]^{3+}; I^-$, have a higher energy than the initial ion pair, they are transient in nature and are detectable only by time-resolved spectroscopic methods [165]. The quantum yield is close to zero as long as there are no scavenging reactions, e.g. catalyzed reduction of water, reductive decomposition of ligands, oxidation of Co(II) by dioxygen, which may follow the formation of successor ion pairs [69, 166].

Further examples of the formation of persistent photoproducts by irradiation into IPCT bands are listed in Table 6.

4.2 Photoreactions Induced by Single-Ion Excitation

An alternative route for IPCT excitation is the excitation of only one partner in the ion pair. This is of particular interest with respect to spectral sensitization since the long-wavelength absorption of ions, e.g. dyestuff ions, can be exploited. The excitation of one partner, here A^+, in the ion pair $(A^+; B^-)$ leads to the ion pair $(^*A^+; B^-)$. If the formation of this ion pair is followed by fast electron transfer, the radical pair $(A^\cdot; B^\cdot)$ is generated. For the subsequent reactions, the same kinetic model as discussed in Sect. 4.1. can be used, although the radical pair $(A^\cdot; B^\cdot)$ has not necessarily to be identical with the successor pair formed by IPCT excitation, see Scheme 3.

Since the direct excitation of A^+ corresponds to an allowed transition, the lifetime of $^*A^+$ is within the ns region. Therefore, processes, such as cage escape, exchange of $^*A^+$ by A^+ or annihilation by reaction with already existing $A^\cdot$, play a minor role. Heavy-atom effects, e.g. in ion pairs containing iodide ions [177], may significantly reduce luminescence lifetimes.

Intersystem-crossing processes (ISC) compete with electron transfer and luminescence and radiationless deactivation, respectively. ISC leads to generation of an ion pair $(^*A'^+; B^-)$ which is distinguished from $(^*A^+; B^-)$ by its spin state. The luminescencc lifetime τ'_{IP} is much longer than τ_{IP}, and other processes become more important. For the irreversible formation of products, the rate of electron transfer k'_{et} is of crucial importance. Although the ion pair $(^*A'^-; B^-)$ has a lower energy than $(^*A^+; B^-)$, the formation of $(A'^\cdot; B'^\cdot)$ may proceed very efficiently if τ'_{IP} is long enough and electron transfer is not too endergonic. In some cases this thermal reaction also proceeds via low-energy intermediates (see Sect. 5.4.).

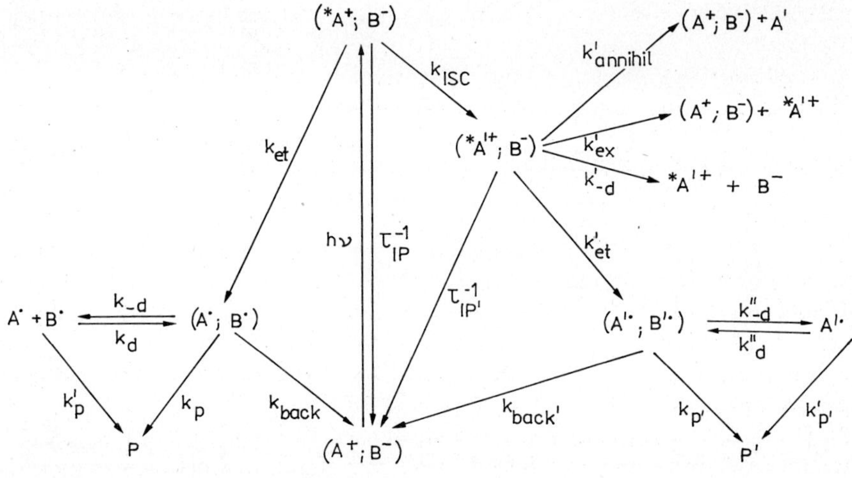

Scheme 3

The successor pair (A′·; B′·) formed by thermal electron transfer does not correlate with the initial ion pair (A^+; B^-) with respect to its spin state. Hence, back electron transfer (k'_{back}) is slow and nonadiabatic. The efficiency of cage escape is given by $\eta'_{-d} = k''_{-d}/(k'_{back} + k''_{-d} + k'_{-et})$.

4.3 Interrelation between Photochemical Reactivity and Thermal Stability

Electron transfer can be achieved not only by photoexcitation but also by thermal excitation. Here, the excitation energy of electron transfer leading to (*A^+; B^-) and (A·; B·), respectively, is the upper limit of the free energy of activation $\Delta G^{\neq}$. The rate constant k_{obs} of product formation can be obtained by using Eq. (15)

$$k_{obs} = k_{et}\Phi_P \tag{15}$$

In Eq. (15) k_{et} and Φ_P are the rate constant of electron transfer and the quantum yield of product formation, respectively. k_{et} can be calculated according to Eq. (16) [161]

$$k_{et} = \nu_n \exp(-\Delta G^{\neq}/RT) \tag{16}$$

Provided the processes under discussion are not too exergonic, the frequency of the nuclear vibration ν_n may be assumed to be $\nu_n = k_BT/h$. An eventually existing nonadiabatic behavior of electron transfer has already been taken into consideration by using Φ_P. Therefrom it follows for k_{obs}

$$k_{obs} = \Phi_P(k_BT/h)\exp(-\Delta G^{\neq}/RT) \tag{17}$$

Table 7. Photoreactions of ion pairs caused by light absorption of one partner (*)

Cation	Anion	Solvent	λ_{irr}[a]	Φ	Ref.
*$[Co(NH_3)_6]^{3+}$	I^-	H_2O	500	0.0015	[168]
$[Co(NH_3)_6]^{3+}$	*$[Pt(mnt)_2]^{2-}$	DMF	347	1.0[d]	[180]
*$[Co(en)_3]^{3+}$	$[Fe(CN)_6]^{4-}$	H_2O, 2M NaCl	647.1	0.9	[46]
*$[Co(en)_3]^{3+//}$	$Cl^{-//}$	solid	254	[b]	[181]
*$[Co(en)_3]^{3+//}$	htc^-	H_2O	451	0.16/7.6[c]	[66]
*$[Co(NH_3)_5(CH_3)_3COO^-]^{2+}$	$(CH_3)_3COO^-$	CH_3OH, CH_2Cl_2	>320	[b]	[182]
*$[Co(NH_3)_5(CH_3)_3COO^-]^{2+//}$	N_3^-	CH_3OH, CH_2Cl_2	>320	[b]	[183]
*$[Co(phen)_3]^{3+}$	$C_2O_4^{2-}$	H_2O	488	0.021	[184, 185]
*$[Co(phen)_2C_2O_4^{2-}]^+$	$C_2O_4^{2-//}$	C_2H_5OH	>320	[b]	[186]
*$[Cu(en)_2]^{2+}$	$[W(CN)_8]^{4-}$	solid	UV	[b]	[187]
*$[Cu(phen)_2]^{2-//}$	BPh_4^-	benzene	405	[b]	[188]
*$[Cu(dmp)_2]^+$	cis-$[Co(IDA)_2]^-$	H_2O, pH 5	454	0.017	[189, 190]
$[Cu(TTAP)]^+$	*$[Zn(TPPS)]^{4-}$	H_2O, acetone	>400	0	[22]
*$[Os(5\text{-}Cl\ phen)_3]^{2+}$	$[Fe(CN)_6]^{4-}$	H_2O	480	0	[191]
*$[Ru(bpy)_3]^{2+}$	$[Mo(CN)_8]^{4-}$	H_2O	452	0	[192, 193]
*$[Ru(bpy)_3]^{2+}$	$[Fe(CN)_6]^{4-//}$	H_2O, CH_3OH	408–480	0	[194]
*$[Ru(bpy)_3]^{2+//}$	$[Ni(mnt)_2]^{2-}$	H_2O, CH_3OH	337/450	0	[151]
*$[Ru(bpy)_3]^{2+//}$	$[CoSiW_{11}O_{39}H_2O]^{6-//}$	H_2O	532	0	[43, 195]
*$[Ru(bpy)_3]^{2+}$	$dtc^{-//}$	CH_3CN	450	[b]	[196]
*$[Ru(bpy)_3]^{2+}$	H_2EDTA^{2-}	H_2O, MV^{2+}	460	0.09	[197]
*$[Ru(bpy)_3]^{2+}$	$S_2O_8^{2-}$	H_2O	355, 450	2.0	[198]
*$[Cr(bpy)_3]^{3+//}$	$C_2O_4^{2-//}$	H_2O	423	1.7	[199]
BV^{2+}	*$[Al(TSCP)]^{3-}$	H_2O		0	[154]
MV^{2+}	*$[Pd(TPPS)]^{4-}$	H_2O		0	[200]
MV^{2+}	*$[Zn(TPPS)]^{4-}$	H_2O		0	[25, 152, 153, 201]
$MV^{2+//}$	*$[Pt(mnt)_2]^{2-}$	DMF	347	1.0[d]	[115]
MV^{2+}	*$[Zn(mnt)_2]^{2-}$	DMSO	308	[b, d]	[114]
MV^{2+}	*EY^{2-}	H_2O, C_2H_5OH	500	0	[202]
*$MB^{+//}$	AMS^-	CH_3OH	640	0	[50, 51]
*$MB^{+//}$	BPh_4^-	H_2O	620	0.1	[203]
*$NSPY^{+//}$	I^-	CH_2Cl_2	353	0	[204, 205]
*ICC^+	BPh_3Bu^-	ethyl acetate	532	0.14	[206]
Ph_2I^+	*RB^{2-}	CH_2Cl_2	>400	[b]	[155]
*$C_6H_5\text{—}N_2^+$	I^-	H_2O	313	0.1	[104]

[a] in nm, [b] product formation not quantitatively measured, [c] autocatalyzed reaction, [d] cage-escape efficiency from flash-photolysis experiments, no permanent product formation // data of similar systems are included in the same reference

and for ΔG^{+}

$$\Delta G^{+} = RT[\ln(K_B T/h) + \ln \Phi_P + \ln \tau] \tag{18}$$

with $\tau = 1/k_{obs}$.

Here, τ is the lifetime of the ion pair, i.e. the time during which the concentration of the ion pair is thermally reduced by the factor 1/e. For a reasonable quantum yield ($\Phi_P = 1\text{–}10^{-6}$) and a lifetime of about one year, ΔG^{+} has to be about 1.2–0.85 eV. Therefrom it follows that the low-energy absorption band of ion pairs has to be in the range of less than 1000–1500 nm in order to achieve a sufficiently long lifetime (of about one year). This means that spectral sensitization with reasonable quantum yields is possible over the entire visible range of the spectrum. However, it should be noted that this model refers only to the most rapid of the possible thermal processes with vertical excitation. In addition, there may exist thermal processes involving vibrationally excited states that require much lower activation barriers. An exact description of the thermal activation barrier of systems exhibiting IPCT character results from the Hush theory [141]. For those systems, faster nonadiabatic processes or processes involving low-energy intermediates or proceeding via precursor ion pairs of different geometry may exist [28].

Strategies to overcome problems arising from thermal instability are discussed in [207].

5 Case Studies of Ion Pair Photoreactions

5.1 Ion Pairs of Diazonium Salts

Aromatic diazonium salts are known as light-sensitive compounds and they have found a wide application in unconventional information recording systems [208]. After light excitation, a diazonium ion can decompose either on an ionic pathway, Eq. (19), or, in the presence of a suitable electron donor X^-, via formation of free radicals, Eq. (20), [209]

$$ArN_2^+ \xrightarrow{h\nu} Ar^+ + N_2 \tag{19}$$

$$ArN_2^+ \xrightarrow{h\nu(X^-)} Ar^{\cdot} + N_2 + X^{\cdot} \tag{20}$$

After ionic decomposition, the aryl cation is usually in its electronic singlet state. However, in the presence of nitrogen substituents in 4-position, the formation of triplet state aryl cations is favoured [210].

The product distribution is a measure of the fraction of radical processes (20).

The formation of ArH is ascribed to the radical process, whereas formation of ArX is regarded as the result of ionic decomposition [211].

Due to the high reduction potential of the electronically excited diazonium ion ($^*E_{red} = 3.91$ V for 4-methyl-benzene diazonium ion in acetonitrile [212]), fast electron transfer reactions are expected even for ions not easily oxidizable, e.g. Cl^- ($E^0 = 2.1$ V in acetonitrile [213]). This rate of electron transfer is, however, limited by diffusion. This limitation can be overcome by formation of ion pairs (ArN_2^+; X^-) in the ground state.

During photolysis of $4\text{-MeC}_6H_4N_2^+BF_4^-$ ($\lambda_{irr} = 313$ nm) in mixtures of dichloromethane and acetonitrile in the presence of oxidizable anions, such as BPh_4^-, Br^-, $HC_2O_4^-$ and Cl^-, an increase of the fraction of ArH among the products has been observed for increasing solvent polarity and descreasing oxidation potential of the anion [212]. This result has been interpreted in terms of ion pairing. Despite the short lifetime of the excited state of diazonium ions ($\tau_0 = 85$ ps for $4\text{-MeC}_6H_4N_2^+$ in acetonitrile [212]) efficient electron transfer quenching of the singlet state has been observed due to strongly negative ΔG^* values. Back electron transfer reduces the efficiency of dediazonation. Thus, a maximum quantum yield of 0.1 has been observed for the formation of toluene in the presence of an excess of bromide ions. A radical process leading to the formation of 4-bromotoluene has been ruled out since the lifetime (300 ns [241]) of the 4-methylphenyldiazo radical formed by electron transfer is much longer than the time the radical pair resides in the solvent cage. In neat dichloromethane, where ion pairing should reach a maximum, a fraction of 60% toluene in the product mixture has been observed during the photolysis of $4\text{-Me-C}_6H_4N_2^+BF_4^-$ in the presence of equimolar amounts of $Et_4N^+Br^-$.

Based on these results, one would expect that ion pairs of the easily oxidizable iodide ($E^0 = 1.33$ V in water [213]) with diazonium ions also react via free radicals. In contrast, the photolysis of aqueous solutions of benzene diazonium iodide led almost exclusively to phenyliodide [104]. It has been demonstrated that the yield of phenyliodide increases with increasing extent of ion pairing. If ion pairing is complete, the quantum yield reaches its maximum which is about 50% of the value measured in the absence of iodide ($\Phi_{N_2} = 0.2$).

It is assumed that the formation of phenyliodide is a result of fast ionic decomposition of the excited diazonium ion and subsequent combination of resulting phenyl cations with iodide ions inside the solvent cage. As a result of electron transfer between iodide and phenyl cations, iodine atoms are formed which may escape and form I_3^- [104]. On the other hand, electron transfer between excited diazonium ions and iodide should result in the formation of benzene which, however, was shown to be of minor importance. In order to offer an explanation for these contradictory results, one has to take into account that the positive charge is essentially localized at the diazonium group [215]. Light absorption leads to charge transfer into the phenyl ring, Eq. (21) [156]

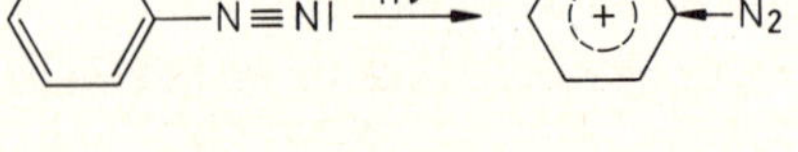

(21)

In case of dynamic quenching by electron transfer, the electron donor should attack at the phenyl ring forming a relatively persistent σ-radical, Eq. (22) [216]

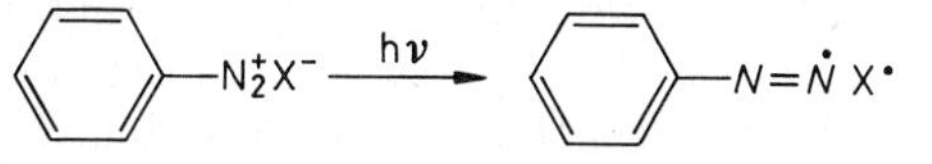

(22)

This aryl diazenyl radical may either regenerate the diazonium cation by back electron transfer or escape from the primary solvent cage to form the corresponding arene.

As shown by X-ray analysis of ion pairs of diazonium salts [215], the anion is positioned at the nitrogen attached to the phenyl ring. Therefore, electron transfer to the excited diazonium ion should proceed through the diazonium group leading to an energetically less favoured radical. Apparently, nucleophilic substitution at the phenyl ring is favoured for this reason. Consequently, during the photolysis of benzene diazonium tetrafluoroborate an increase of the fraction of fluorobenzene is observed if ion pairing progresses. However, the participation of radical intermediates in the substitution as proposed by Sahyun [104] for benzene diazonium iodide has not yet been proven because I_3^- observed by this author could also be the result of dynamic quenching of excited diazonium ions by iodide ions in the bulk solution. From this discussion it follows that the increased quantum yield for the formation of ArH observed by Israel et al. [212] is probably due to kinetic effects during dynamic quenching rather than to the presence of ion pairs. According to the discussion of these authors, the yield of ArH should depend on the rate of the back electron transfer between diazenyl radicals and axidation products of the anions.

The electron transfer rate for ion pairs of diazonium ions is determined by both the relatively high reorganizational energy of the couple ($ArN_2^+/ArN_2^{\cdot}$), $\Delta G^{\neq}(0) = 83\text{–}91\ kJ\ mol^{-1}$ [217], and the driving force ΔG. Since decreasing donor ability of the solvent causes a shift of the reduction potential of cations towards more positive values [144], and, on the other hand, decreasing acceptor strength of the solvents leads to more negative values of the oxidation potential of the anions [59], a change from acetonitrile to dichloromethane (see Tables 2 and 3) should be accompanied by more positive ΔG values and, hence, slower back electron rates leading to an increase of ArH yield. As compared with direct excitation of diazonium ions, ion pairs with IPCT absorption bands are characterized by a different situation. Thus, irradiation into the IPCT band leads to an electron transfer from the anion to the diazonium group which evokes the formation of a relatively persistent σ-radical and oxidized anions. This is in agreement with investigations performed by Vogler [105] on ion pairs of hexacyanoruthenate(II) with 4-methoxybenzenediazonium ions (see Table 6). As far as we are aware no further studies of the photochemistry of ion pairs of diazonium ions with IPCT behavior are known which parallels the low thermal stability of these ion pairs [217].

5.2 Ion Pairs of Diaryliodonium Salts

Light-sensitive diaryliodonium salts are used in information recording materials and serve as initiators in radiation-hardenable layers [218]. Direct irradiation of UV absorbing diphenyliodonium ions leads to the generation of iodoaryl cation radicals and aryl radicals, Eq. (23), from either triplet [104] or nonspectroscopic singlet states [219]

$$Ar_2I^+ \xrightarrow{h\nu} Ar^{\cdot} + ArI^+ \tag{23}$$

The reaction of the excited triplet state with a reducing agent as well as reduction of iodonium ions in its electronic ground state leads to a diaryliodyl radical, $Ar_2I^{\cdot}$, for which a lifetime of 20 μs has been measured [219]. However, reduction of diphenyliodonium ions by excited anthracene shortens the lifetime of the diaryliodyl radical to 250 ps [220].

Ion pairs of Ph_2I^+ with easily oxidizable anions, such as iodide [103, 104] and cyanometallates [102], exhibit additional long-wavelength IPCT absorption bands. The position of the absorption maximum is only little effected by substituents in the phenyl ring [102, 103] which is interpreted in terms of localization of the positive charge at the iodine. These additional absorption bands render ion pairs of diaryliodonium salts suitable for spectral sensitization. The influence of both the oxidation potential of the anion and the solvent on the energy of the IPCT band has been demonstrated [102–104].

Irradiation of aqueous solutions of diphenyliodonium iodide in the range of the IPCT band (λ_{irr} = 355 nm) leads exclusively to the formation of phenyliodide [104, 221] with a quantum yield $\Phi = 0.05$. Interestingly, the charge-transfer excited state has a higher energy than the (nonspectroscopic) excited singlet state of the cation. It has been shown by flash-photolysis that within 200 ps after charge-transfer excitation the system relaxes into the dissociative singlet state leading to phenyliodide and phenyl cations. In analogy to benzenediazonium iodide (see Sect. 5.1), the latter may react with iodide in the cage to form phenyliodide. The formation of phenyl radicals may be ruled out since, in contrast to other iodonium salts, diphenyliodonium iodide does not initiate the photopolymerization of vinyl chloride [222].

On the other hand, for the more easily oxidizable cyano complexes of e.g. Ru(II), Mo(IV) and W(IV) photoinduced electron transfer caused by irradiation into the IPCT bands results in the oxidation of the anion ($\Phi = 0.5$ for $[Mo(CN)_8]^{4-}$ in methanol [102]) and formation of diphenyliodyl radicals. The latter initiate chain reactions in methanol solution (Scheme 4) with quantum yields for the benzene formation clearly exceeding unity [102].

The possibilities of long-wavelength spectral sensitization of photoreduction of diphenyliodonium ions by ion pairing with easily oxidizable anions, e.g. $[Fe(CN_6]^{4-}$ (E^0 = 0.36 V vs NHE [223]) are limited by thermal redox reactions that occur at room temperature [102]. This disadvantage can be overcome by choosing ionic dyes as counterions which have an oxidation potential high enough to provide thermal stability in their ground states but which may reduce diphenyliodonium

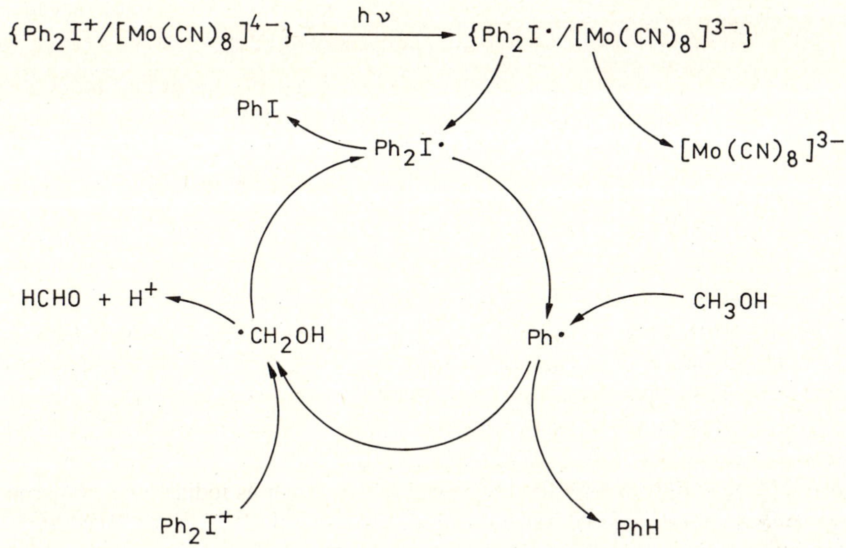

Scheme 4

ions when present in their excited states. These conditions are met e.g. for xanthene dyes such as rose bengale (RB^{2-}, $E^0 = 0.86$ V [155]). Ion pairs of diphenyliodonium ions with RB^{2-} in dichloromethane are readily bleached upon irradiation into the absorption bands of the dye, whereas in methanol, where no ion pairs are formed, no oxidative bleaching of dye molecules takes place. In this reaction, the excited singlet state of the dye is oxidately quenched by Ph_2I^+. Due to rapid decomposition of diphenyliodyl radicals, *O*-arylated products of the dye are formed in the solvent cage. Photoreduction of diphenyliodonium ions in ion pairs with RB^{2-} takes place exclusively at the phenolic oxygen of the dye, whereas Ph_2I^+ bound to the carboxylic group is not reduced [155].

5.3 Ion Pairs of Cobalt(III) Amine and Diimine Complexes

Cobalt(III) complexes with amine and diimine ligands are known to be extremely inert. In contrast, the corresponding cobalt(II) complexes are labile and decompose very rapidly in donor solvents such as water forming Co^{2+}_{aq} and free ligands [224] unless the decomposition is prevented by a cage structure of the ligands, e.g. for sepulchrate complexes [225].

Irradiation into charge-transfer bands (approx. 254 nm) of cobalt(III) complexes leads to photoreduction in aqueous solution, whereas photosubstitution reactions are preferred for excitation of ligand-field bands [167]. Apparently, the energy is not sufficient in order to cause photooxidation of the solvent and of ligands in the first coordination sphere, respectively. For coordination of more easily oxidizable anions in the second coordination sphere, i.e. ion pairing, one would

Table 8. Reduction potentials of cobalt(III) complexes leading to the corresponding high-spin (h.s.) and low-spin (l.s.) cobalt(II) complexes in aqueous solution [226]

	$L = NH_3$	en	sep	tacn	phen	bpy
E° (h.s.) [V]	+0.06	−0.21	−0.30	−0.41	+0.39	+0.31
E° (l.s.) [V]	−0.76	−0.68	−0.65	−0.51	−0.09	+0.10

expect a shift of photosensitivity towards longer wavelength. IPCT behavior has been shown for a series of cobalt(III) complexes (see Table 4). A particularity of most cationic cobalt(III) complexes is that the reduction potentials measured by electrochemical methods refer to the transition Co(III) (t_{2g}^6) to high-spin Co(II) ($t_{2g}^5e_g^2$). On the other hand, the primary product formed by light-induced electron transfer in ion pairs is Co(II) ($t_{2g}^6e_g^1$). The difference between high-spin and low-spin redox potentials is about 0.1 to 0.8 V (Table 8). For this reason the IPCT absorption maxima of cobalt(III) complexes are shifted towards shorter wavelengths as compared with the corresponding ruthenium(III) complexes which have similar (electrochemically measured) reduction potentials.

Irreversible formation of photoproducts upon irradiation into IPCT bands is observed if either the resulting cobalt(II) complexes are labile [56, 162, 163, 167–169] or the oxidation products of the anions decomposes as observed for oxalate [57, 171] and tetraphenylborate [61, 71] (see Sect. 5.4. and 5.5.).

If these requirements are not met, e.g. for $[Co(sep)]^{3+}$; I^-, photolysis does not result in persistent products, although photoproducts may be detected as transient intermediates [165, 166, 173].

Again, back electron transfer in the successor pair, which competes with cage escape and subsequent decomposition, reduces the efficiency of IPCT excitation. Since the latter process is limited by the rate of dielectric relaxation of the solvent (see Sect. 4.1), the quantum yield increases with decreasing radii and charges of the photoproducts (Table 5). The conversion of the spin state of low-spin cobalt(II) complexes generated by photolysis of ion pairs into thermodynamically more stable high-spin complexes has apparently no influence on the back electron transfer in the successor pair. This indicates that the latter is slow as compared with cage escape. On the other hand, it has been found that for ion pairs of cobalt(III) complexes in solid state there is an influence of spin state conversion upon the quantum yield [227, 228].

The possibility of spectral sensitization of cobalt(III) complexes by forming ion pairs with IPCT behavior is limited by the decreasing thermal stability of the associates. Thus, the ion pair $[Co(en)_3]^{3+}$; $[Fe(CN)_6]^{4-}$ ($\bar{\nu}_{CT} = 22{,}200\ cm^{-1}$) undergoes slow thermal electron transfer at room temperature [64].

An alternative is provided by irradiation into the long-wavelength, though less intense ligand-field bands. In case of ammine and diaminoethane complexes fast ISC processes may lead to an excited quintet state of cobalt(III) ($t_{2g}^4e_g^2$) which provides a favourable configuration for the reduction to the thermodynamically more stable high-spin state ($t_{2g}^5e_g^2$) of the corresponding cobalt(II) complexes

[46, 229]. Electron transfer in the successor pair is then spin-forbidden and slow (see Scheme 3).

The quantum yield for the photoreaction of ion pairs of $[Co(en)_3]^{3+}$ with easily oxidizable hexacyanoferrate(II) ($E^0 = 0.36$ V [223]) reaches unity upon irradiation into the ligand-field bands (488 and 647 nm, respectively). However, due to high charges of the partners cage escape of the successor pair is slow. Therefore, cobalt(III) is regenerated after substitution of one diaminoethane ligand and back electron transfer in the resulting dinuclear complex [46]. If optically active (+) $[Co(en)_3]^{3+}$ is used, only partial racemization is observed. The quantum yield is reduced to values between 0.38 and 0.17 at high ionic strengths, although ion pairs still absorb about 70% of the incident light. The dependence of the quantum yield on the concentration of hexacyanoferrate(II) was interpreted by Langford [46] as diffusion-controlled quenching of excited cobalt(III) in the ion pair by free $[Fe(CN)_6]^{4-}$. We were able, however, to show by multiparameter regressional analysis of the data published by Langford that there is an dependence of the quantum yield on the concentration of free $[Co(en)_3]^{3+}$ rather than on the concentration of hexacyanoferrate(II). This suggest an exchange of excited cobalt(III) complexes in the ion pair by free Co(III) from the bulk solution to be the reason for the reduction of quantum yield (see Scheme 3).

Since the quintet state of $[Co(NH_3)_6]^{3+}$ (and probably also that of the complex $[Co(en)_3]^{3+}$) lies less than 13000 cm^{-1} above the ground state [229], electron transfer from poorly oxidizable anions is slow and quantum yields become smaller (see Table 9).

Due to the high ligand-field strength of the dimine ligands bpy and phen, the quintet state of the corresponding cobalt(III) complexes has an unfavourable energy. After ligand-field excitation the triplet state ($t_{2g}^5 e_g^1$) is populated by ISC, and in analogy to IPCT excitation, low-spin cobalt(II) is formed subsequent to electron transfer from the anion. Since back electron transfer is not spin-forbidden it should proceed rapidly. The quantum yields of photoreduction of $[Co(phen)_3]^{3+}$ in the presence of a large excess of oxalate is $\Phi = 0.03$ upon irradiation into the ligand-field band (488 nm) [185]. It resembles the value obtained during irradiation into the IPCT bands (313 nm, $\Phi = 0.06$) [57].

If the photolysis of ion pairs of cobalt(III) is carried out in the presence of potential ligands, e.g. $EDTA^{2-}$ [172], 1-nitroso-2-naphthol [230], pyridyl azonaphthol [231], dtc^- [232], htc^- [66] or S^{2-} [233], the cobalt(II) complexes of which are

Table 9. Quantum yields for formation of permanent products by longwavelength (550–600 nm) photolysis of ion pairs $[CoL_n]^{3+}$; B^- in aqueous solution

B^-	$E°(B^\cdot/B^-)$ [V]	L	Φ	Product	Ref.
$[Fe(CN)_6]^{4-}$	0.36	en	1.0	$[Cl(en)_2Co{-}NC{-}Fe(CN)_5]^{2-}$	[46]
htc	—	en/NH_3	0.16/0.20[a]	$[Co(htc)_3]$	[66]
I^-	1.33	NH_3	0.0015	$Co^{2+}_{aq.}$	[168]
$C_2O_4^{2-}$	2.1	en/NH_3	$<10^{-4}$	—	[57, 185]

[a] corrected with respect to catalyzed secondary reactions

able to reduce initial cobalt(III) complexes, photoinduced catalytic substitution of kinetically inert cobalt(III) complexes with quantum yield greater than 1 may be accomplished (Scheme 5).

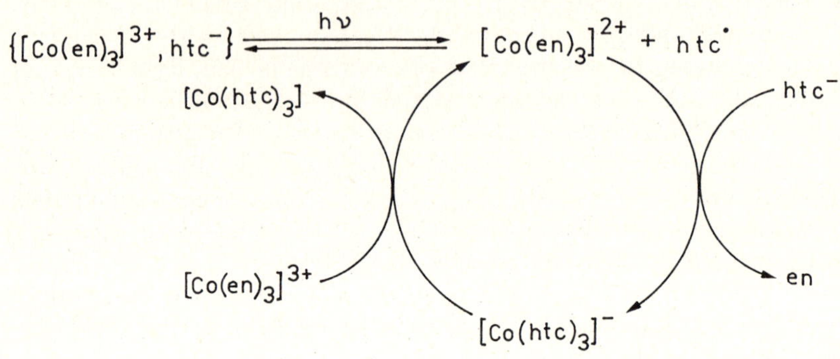

Scheme 5

Those photoinduced catalytic processes can be applied in unconventional photographic processes [231, 233–238].

A third possibility of spectral sensitization of photoreduction of cationic cobalt(III) complexes is provided by the excitation of the counterion as it has been demonstrated for the ion pair $[Co(NH_3)_6]^{3+}$; $[Pt(mnt)_2]^{2-}$ [180].

Very high yields of $[Pt(mnt)_2]^-$ ($\Phi = 1, \lambda_{irr} = 347$ nm) have been measured by flash photolysis in DMF. However, unlike in water the resulting cobalt(II) complex does not decompose rapidly. Therefore, slow back electron transfer ($k = 6\,M^{-1}\,s^{-1}$) leads to regeneration of the starting compounds.

Attempts to sensitize the photoreduction of cobalt(III) amine complexes by ion pairing with anionic dyes, e.g. rose bengale and eosin, also led to relatively high yields of Co(II) but a dynamic mechanism cannot be ruled out [57].

Interestingly, the photoreduction of the ion pair $[Co(NH_3)_6]^{3+}$; Cl^- can be sensitized by naphthalene, whereas sensitization of the free complex ion has failed [239].

5.4 Ion Pairs Containing Tetraarylborates

Insoluble salts of easily reducible salts of alkyl and arylborates, R_4B^-, form soluble products upon irradiation. This reaction has been used as a basis for photolithographic systems, and a large number of alkyl and arylborates with more than 400 different cations have been claimed as initiators for photolithographic uses [240].

Due to its large volume ($r = 460$ pm [241]), sodium tetraphenylborate, which shows only little photosensitivity [242], has a low tendency of forming ion pairs with cations [34, 61, 71, 73, 144, 241]. Therefore, contact ion pairs are formed in nonpolar media only. However, if the steric situation allows a π–π interaction between planar cations, such as diazonium ions, methyl viologen and planar Ni(II) complexes, and a phenyl ring of the tetraphenyl borate, a dramatic increase of ion pairing may be observed [52, 53, 243, 244].

Despite the low oxidation potential of the tetraphenylborate ion ($E^0 = 1.44$ V in acetonitrile [245]) the IPCT absorption bands of ion pairs with easily reducible cations, e.g. methyl viologen [53, 118] and cobalt(III) amine complexes [61, 71], have a fairly high energy as compared with those of the corresponding iodides ($E^0 = 1.2$ V in acetonitrile [213]), which points toward a high Franck-Condon energy.

Upon irradiation into the IPCT band, photoinduced electron transfer leads to the formation of tetraphenyl boryl radicals, $R_4B^{\cdot}$. Apparently, decomposition of the latter favours the irreversible formation of products.

It is noteworthy that various decomposition modes have been observed for the tetraphenyl boryl radical. While IPCT excitation of ion pairs of tetraphenylborate with MV^{2+} [118] and diazonium cations [211, 212] as well as electrochemically generated $R_4B^{\cdot}$ radicals [245] favour the formation of substituted cyclohexadienyl radicals with diphenyl as the ultimate product [246], photolysis of ion pairs with cobalt(III) and copper(II) complexes lead to phenyl radicals [61, 188, 227]. Presumably, this is due a catalytic influence of the metal complexes, eventually via intermediate phenyl complexes. This is also reflected by the higher quantum yields in case of the cobalt(III) complex ($\Phi_{436} = 0.52$ [170]) as compared with ion pairs of methyl viologen ($\Phi_{436} = 0.008$ [118]).

In contrast to tetraphenylborate, tetra[3,5-bis(trifluoromethyl)phenyl]borate, $TFPB^-$, which also undergoes IPCT interactions with MV^{2+}, is stable upon photoinduced electron transfer and the detection of stable products has failed in dimethoxyethane solution. On the other hand, $MV^{+\cdot}$ and $TFPB^{\cdot}$ radicals have been detected in the solid state at low temperature. Upon warming or dissolution of the photolyzed crystals, the starting compounds are regenerated by back electron transfer [125, 176].

Ion pairs of $TFPB^-$ provide the first example for fluorescence at room temperature originating from an IPCT state [159]. It has been shown that the fluorescence can be assigned to the successor ion pair. This is a result of the low tendency of this ion pair to undergo cage escape [52].

Ion pairs of alkyl and arylborates with cationic coordination compounds, organometallic compounds, organic onium compounds and cationic dyes, respectively, undergo also photoredox reactions when irradiated into the long-wavelength absorption bands of the cations [203, 206, 240, 247–250]. It is assumed that short-lived singlet states of the dyes are quenched by R_4B^- in the ion pair. Among the borates, triphenyl-*n*-butyl borate shows the highest efficiency in ion pairs with cyanine dyes which is explained in terms of the higher stability of *n*-butyl radicals [250]. According to recent studies, the fluorescence of cyanine dyes is not quenched by ion pairing with Ph_3BuB^-. In the presence of tri(*n*-butyl)-stannane as a radical scavenger, quantum yields up to 0.14 were measured [206].

5.5 Ion Pairs Containing Carboxylates

Carboxylate ions $RCOO^-$ have been recommended as sacrificial compounds in photochemical processes for solar energy conversion [84, 171, 121] and for information recording materials [251]. Carboxylates themselves absorb light only at very short wavelengths and exhibit low photosensitivity [252]. Since the corresponding carboxylic acids are rather weak, free carboxylate ions exist in higher concentrations only above pH 4.

If the carboxylates do not contain easily oxidizable substituents, the highest occupied molecular orbital is mainly determined by the $p_z(\sigma)$ orbital of the carboxylic group as shown by MNDO calculations [253]. Therefore, the one-electron oxidation potential of simple carboxylate anions should be essentially independent of the substituents. This assumption has been confirmed by recent studies of luminescence quenching of $*[UO_2F_2]$ in aqueous solution [254], Fig. 4. This is also in agreement with the observation that the absorption maximum of IPCT bands of ion pairs of the type $[Co(NH_3)_6]^{3+}$; $RCOO^-$ is only little dependence on R [57].

Due to fast decomposition of the resulting radicals, one-electron oxidation potentials are not accessible by electrochemical measurements. However, estimations on the basis of other methods yielded values between 2.0–2.4 V (in water vs NHE) [255–257].

$$RCOO^- \xrightarrow{-e^-} RCOO^{\cdot} \tag{24}$$

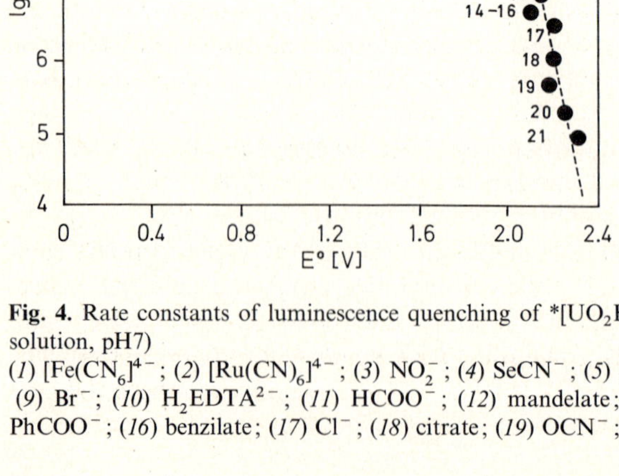

Fig. 4. Rate constants of luminescence quenching of $*[UO_2F_2]$ by various anions B^- (aqueous solution, pH7)
(*1*) $[Fe(CN_6]^{4-}$; (*2*) $[Ru(CN)_6]^{4-}$; (*3*) NO_2^-; (*4*) $SeCN^-$; (*5*) I^-; (*6*) $S_2O_3^{2-}$; (*7*) N_3^-; (*8*) SCN^-; (*9*) Br^-; (*10*) H_2EDTA^{2-}; (*11*) $HCOO^-$; (*12*) mandelate; (*13*) ascorbate; (*14*) $C_2O_4^{2-}$; (*15*) $PhCOO^-$; (*16*) benzilate; (*17*) Cl^-; (*18*) citrate; (*19*) OCN^-; (*20*) pyruvate; (*21*) CH_3COO^-

The thus estimated data for oxidation potentials of carboxylate ions provide a firm basis for the interpretation of processes which involve an electron transfer from the carboxylate to an acceptor. This process is duistinguished by a high reorganizational energy, $\Delta G^{+}(0) = 350$ kJ mol^{-1} [256].

However, if the process is strongly endergonic and, therefore, slow in the timescale of nuclear vibrations, other pathways appear to be possible, especially for dicarboxylates such as oxalate. Those may require lower potentials and lower reorganizational energies, if the formation of high-energy intermediates such as acyloxy radicals, RCOO·, is avoided. As can be seen from Scheme 6, oxidation of oxalate anion to the corresponding acyloxy radical is by 1.3 eV less favourable than the oxidation to CO_2 and $\cdot CO_2^-$.

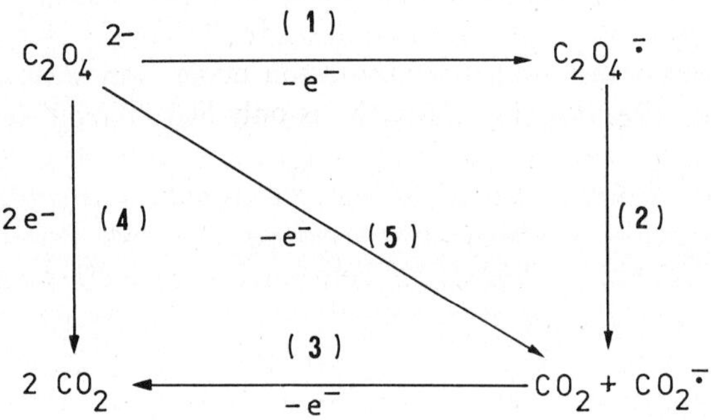

Scheme 6

In Scheme 6, the free energies for the reactions (1), (3), and (4) have the values $\Delta G_1 = 2.1$ eV [257], $\Delta G_3 = -1.9$ eV [258], and $\Delta G_4 = 2x(-0.55)$ eV $= -1.1$ eV [223], respectively.

With $\Delta G_5 = \Delta G_4 - \Delta G_3$ the free energy change for the process (5) may be estimated to be 0.8 eV.

The oxidation potentials of H_2EDTA^{2-} ($E^0 = 1.5$ V [259]) and the monoprotonated ascorbate ($E^0 = 0.71$ V [260]), which have been obtained from the rate of quenching of $*[Ru(bpy)_3]^{2+}$ and from kinetic data on the thermal reduction of ruthenium(III) amine complexes, respectively, are also well below the abovementioned range of 2.0–2.4 V.

Different decomposition modes have also been detected for acetate ions [261] where rapidly decarboxylating û-radicals ($k_{decarbox} = 1.6 \times 10^9$ s^{-1}) and relatively persistent π-radicals were observed.

Ion pairs of cationic cobalt(III) complexes [57, 70, 171], ruthenium(III) complexes [84] and MV^{2+} [119–123, 178, 179] with carboxylates, such as oxalate H_2EDTA^{2-}, citrate^{n-} and aliphatic dicarboxylates (see Tables 4 and 6), also exhibit IPCT absorption bands in the short-wavelength spectroscopic region. The relatively high energy of these IPCT transition is a consequence of both the high oxidation potential and reorganizational energy of carboxylates. A considerable

Table 10. Comparison of calculated and experimental cage-escape efficiencies for ion pairs A^+; $C_2O_4^{2-}$ in aqueous solution

A^+	$r_{A\cdot}$ [pm][a]	$\eta_{-d,calc.}$[b,c]	$\eta_{-d,calc.}$[b,d]	$\eta_{-d,exp.}$[e]	Ref.
$[Co(sep)]^{3+}$	460	0.029	0.16	0.145	[171]
$[Ru(NH_3)_5py]^{3+}$	335	0.035	0.24	0.175	[84]
MV^{2+}	500	0.028	0.14	0.12	[121]

[a] from Ref. [164], [b] $\eta_{-d} = k_{-d}/(k_{back} + k_{-d})$, see Eq. (5) and Sect. 4.1., [c] escape of $C_2O_4^{-\cdot}$ (r = 200 pm, Ref. [199]), [d] escape of CO_2 (r = 130 pm, estimated), [e] equal one half of the quantum yield of A·

bathochromic shift of these IPCT absorption bands by ion pairing with cations having more positive reduction potentials is not possible because of decreasing thermal stability of these ion pairs.

Upon irradiation into the IPCT bands of ion pairs A^+; $RCOO^-$, an acyloxy radical RCOO· is formed, which decomposes with the liberation of carbon dioxide and a radical R· [262, 263]. In the case of oxalate [84, 121, 171], benzilate [120] and citrate (at pH 4) [70], the resulting radicals R· are able to reduce a further equivalent of A^+. Therefore, the quantum yield of these systems, with respect to formation of A, reaches values of up to the double of the cage escape efficiency η_{-d} of the primary radical pair. Cage escape efficiencies of ion pairs of oxalate calculated in analogy to those given in Table 5 are summarized in Table 10.

It should be noted that the calculated data are about 4–5 times smaller than the experimental ones. However, if one assumes rapid decomposition of oxalate monoanion radicals into CO_2^- and carbon dioxide and calculates cage escape efficiencies based on carbon dioxide, good agreement with the data listed in Table 10 is noted.

The conclusion that the oxalate monoanion radical $C_2O_4^{-\cdot}$ decomposes within the time domain of 10^{-10} s contrasts, however, with the results obtained by pulse radiolysis [121, 264] and EPR spin trapping experiments [265] which suggest a lifetime in the microsecond region. It was also demonstrated that free oxalate monoanion radicals are not formed in the photolysis of ion pairs with methyl viologen [121] and $[Co(sep)]^{3+}$ [57]. On the other hand, one would expect that the concentration of the reaction product of the cation, due to back electron transfer (bulk recombination), does not exceed a certain limit. Apparently, the decomposition of oxalate monoanion radicals in the solvent cage is accelerated by interaction with the still oppositely charged reduction product of the cation.

Because of the high energy of IPCT bands of ion pairs with carboxylates it appears to be advantageous for spectral sensitization to use long-wavelength absorption bands of the cations which are redox-inactive in the free cation. Thus, ion pairs of cobalt(III) diimine complexes with oxalate anions undergo photoredox reactions when irradiated into the long-wavelength ligand-field bands [184–186], whereas $[Co(en)_3]^{3+}$ is not reduced under the same conditions [185].

Photoinduced electron transfer upon irradiation into the absorption band of the cation has also been demonstrated for diazonium oxalates by EPR spin trapping [212].

In a detailed study, Espension et al. [199] discussed the quenching of the excited ^2E states of chromium(III) polypyridyl complexes $[CrL_3]^{3+}$ by ion pairing with oxalate. Extremely long luminescence lifetimes of these complexes allow the study of slow electron transfer processes. The quantum yield with respect to formation of the corresponding chromium(II) complex reaches the maximum value of $\Phi = 2$ for L = 5-chlorophenanthroline. Due to annihilation of the excited state by Cr(II) complexes formed during the photoreaction, the quantum yield is reduced to values between 0.4 and 0.6 for complexes with L = 4,4′-dimethylbipyridine.

The high quantum yields in these systems are due to the following reasons: at first, the back electron transfer is spin-forbidden (k'_{back}, Scheme 3) and, secondly, the formation of thermodynamically less favoured oxalate monoanion radicals is avoided. A reasonable explanation of the kinetic findings is possible only by assuming direct decomposition of oxidized oxalate into carbon dioxide and $CO_2^{-\cdot}$ (see Scheme 6, path (5)). In contrast to ion pairs with IPCT absorptions [171], no photoreactivity of chromium(III) ion pairs with oxalate was observed at pH 4, i.e. in a medium where only monoprotonated oxalate, $C_2O_4H^-$, is present. A further interesting example for ion-pair effects in the photochemistry of coordination compounds is the photoreaction in a system containing $[Ru(bpy)_3]^{2+}$, H_2EDTA^{2-} and MV^{2-} [197]. In the absence of H_2EDTA^{2-}, photoexcited $*[Ru(bpy)_3]^{2+}$ is oxidized by MV^{2+} forming $[Ru(bpy)_3]^{3+}$ and $MV^{+\cdot}$. Due to back electron transfer, the quantum yield for the formation of persistent products is $\Phi < 10^{-5}$. No change in luminescence has been observed upon addition of H_2EDTA^{2-}. Therefore, electron transfer within the ion pair can be ruled out. However, if the excited state of $[Ru(bpy)_3]^{2+}$ is oxidized by MV^{2+}, the resulting Ru(III) complex may be reduced by the counterion H_2EDTA^{2-}. The oxidized form of the latter decomposes rapidly. Hence, an oxidation of $MV^{+\cdot}$ by Ru(III) is prevented ($\Phi = 0.09$).

5.6 Miscellaneous Ion Pairs

The irreversible formation of products by photoinduced electron transfer in ion pairs has also been observed for triphenylsulfonium iodide [132]. Probably, this reaction proceeds according to a mechanism similar to the one observed for diphenyliodonium iodide (see Sect. 5.2.).

Dimerization of the oxidation product of the oxidized anion has been found to be responsible for the irreversible IPCT photoreaction of MV^{2+}; dimethyldithiophosphate ion pairs [124] and for the formation of $[Ru(bpy)_3]^{3+}$ from ion pairs with dithiocarbamate [196].

Ion-pair effects have also been described for the photolysis of $[Ru(bpy)_3]^{2+}/S_2O_8^{2-}$ in aqueous acetonitrile [198]. Here, both static and dynamic processes of equal efficiency ($\Phi = 2$) leading to Ru(III) have been observed.

Ion pairs of vanadium(V) have been recommended as initiators for the photopolymerization of methylmethacrylate [266]. Interestingly, the corresponding free vanadium complexes have failed to initiate radical polymerization.

Fast aquation was found to prevent cage recombination of the labile photoproducts $[Ru(NH_3)_5Cl]^+$ [86] and $[Rh(bpy)_3]^{2+}$ [88] obtained by irradiation into the IPCT bands ion pairs of $[Fe(CN)_6]^{4-}$.

Static quenching [161] of luminescence of excited ions, e.g. $*[Ru(bpy)_3]^{2+}$, $*[Eu(2.2.1)]^{2+}$ and ionic metallo porphyrins (see Table 7), due to ion pairing is a very common phenomenon. Here, ISC processes leading to long-lived excited states are suppressed and their photoreactions are quenched.

Similar effects have been observed for ion pairs of the iodide with 4-cyanotrimethylanilinium cations [133], crystal violet [134] and substituted azastilbenes [205]. On the other hand, the lack of formation of $MV^{+\cdot}$ during photolysis of ion pairs $MV^{2+}; I^-$ is ascribed to a heavy-atom effect [117].

If photoinduced electron transfer is not possible for energetic reasons, e.g. for $*[Os(phen)_3]^{2+}$, ion pairing may even lead to an increase of excited states lifetimes due to the polarizing effect of the counterion [24].

Recently, several papers have been published dealing with the photochemistry of ion pairs of metal carbonyl complexes, e.g. $[Co(CO)_4]^-$ [74, 76, 77, 127, 174]. In some cases, these ion pairs are thermally stable and exhibit IPCT absorption bands at very low energies.

17-electron metal carbonyls are formed when ion pairs of $[Co(CO)_4]^-$ are photolyzed

$$A^+; [Co(CO)_4]^- \xrightarrow{h\nu_{IPCT}} A^\cdot + [Co(CO)_4]^\cdot \quad (25)$$

This photoreaction is almost completely reversible. Only upon addition of phosphines and phospites, respectively, which undergo fast substitution reaction with intermediate carbonyl metal radical, Eq. (26), back electron transfer can be prevented

$$[Co(CO)_4]^\cdot + L \rightarrow [Co(CO)_3L]^\cdot + CO \quad (26)$$

Depending on the nature of the ligand L, $[Co(CO)_3L]^\cdot$ may react by dimerization, substitution, and disproportionation, respectively, Eqs. (27)–(29)

$$2[Co(CO)_3L]^\cdot \rightarrow [Co_2(CO)_6L_2] \quad (27)$$

$$L = PPh_3$$

$$[Co(CO)_3L]^\cdot + A^\cdot \rightarrow [Co(CO)_3L]^- + A^+ \quad (28)$$

$$L = P(OPh)_3$$

$$[Co(CO)_3L]^\cdot + A^+; [Co(CO)_4]^- + L \rightarrow [Co(CO)_3L_2]^+; [Co(CO)_4]^- + A^\cdot \quad (29)$$

$$L = PBu_3$$

Particularly, the disproportionation appears to be an efficient way for generating products. Thus, a quantum yield of $\Phi = 0.44$ has been measured for the IPCT photolysis ($\lambda_{irr} = 550$ nm, THF) of the ion pair of the quinolinium cation with $[Co(CO)_4]^-$ in the presence of 0.1 M tributylphosphine [174], whereas a quantum yield of only 0.02 was found for the photolysis of the structurally similar ion pair $DMPBY^{2+}$; $[Co(CO)_4]^-$ ($\lambda_{irr} = 514$ nm, CH_2Cl_2) in presence of 0.1 M triphenylphosphine [127].

6 Conclusions and Perspectives

As compared with separate ions, ion pairs may exhibit a variety of novel interesting spectroscopic and photochemical properties. The appearance of additional ion-pair charge-transfer (IPCT) absorption bands, which correspond to an electron transfer from the anion to the cation, has been demonstrated for a large number of ion pairs. The converse process, i.e. the transfer of an electron from a donor cation to an acceptor anion has not yet been observed in solution. Favourable conditions for the detection of this "inverse" IPCT behavior are offered by metal complexes since they are very often characterized by a number of well-defined stable oxidation forms.

The formation of ion pairs also provides new ways for photoinduced electron-transfer reactions. The suppression of back electron transfer appears to be the crucial problem to be solved in order to achieve high yields of products. The highest yields are observed if the spin state of the ion pair is changed after light excitation but prior to electron transfer. A further requirement to be met for obtaining high efficiencies of product formation is rapid irreversible decomposition of the successor pair.

Many of the investigated ion pair systems consist of relatively small symmetric ions attracted by mainly electrostatic forces. However, if the participating ions have a more complex structure, then there is the possibility of additional stabilization of certain conformations of ion pairs. The thus created stereoselectivity of ion-pair formation may effect the photoreactions of these ion pairs. The study of these effects is only at its very beginning. It may, however, not only provide interesting insights into mechanistic details of the photochemistry of ion pairs but also open new fields of application to unconventional photographic processes, selective syntheses, and modelling of biochemical processes.

7 References

1. Fox MA, Chanon M (eds) (1988) Photoinduced electron transfer, Elsevier, Amsterdam
2. Hennig H, Rehorek D, Archer RD (1985) Coord. Chem. Rev. 61: 1
3. Hennig H, Rehorek D (1988) Photochemische und photokatalytische Reaktionen von Koordinationsverbindungen, Teubner, Stuttgart
4. Chibisov AK (1984) Prog. React. Kinet. 13: 1
5. Hennig H, Thomas P, Wagener R, Ackermann M, Benedix R, Rehorek D (1981) J. Signal Aufzeichnungsmat. 9: 269

6. Werner A (1912) Ber. Dt. Chem. Gesellsch. 45: 121
7. Balzani V, Sabbatini N, Scandola F (1986) Chem. Rev. 86: 319
8. Palmer RA, Carter RC, Dyer RB, Ghirardelli RG, Metcalf DH (1987) Understanding molecular properties, Reidel, Dordrecht, p 359
9. Colquhoun HM, Stoddart JF, Williams DJ (1986) Angew. Chem. Int. Ed. Engl. 25: 487
10. Eaton, DR (1988) Rev. Chem. Intermed. 9: 201
11. Nekipelov VM, Zamaraev KI (1985) Coord. Chem. Rev. 61: 185
12. Brown DB (ed.) (1980) Mixed-valence compounds, D. Reidel, Dordrecht
13. Robin MB, Day P (1967) Adv. Inorg. Chem. Radiochem. 10: 247
14. Benesi HA, Hildebrand JH (1949) J. Am. Chem. Soc. 71: 2703
15. Drago RS, Rose NJ (1959) J. Am. Chem. Soc. 81: 6138
16. Darensbourg MY (1985) Prog. Inorg. Chem. 33: 221
17. Fanali S, Ossicini L, Sinibaldi M (1987) Chromatographia 23: 811
18. Masuda Y, Yamatera H (1988) J. Phys. Chem. 92: 2067
19. Borovnikov MS, Geosdovski GN, Rybakov VA, Tarasov BP (1982) Zhurn. Obshkh. Khim. 52: 331
20. Miyoshi K, Oh CE, Yoneda H (1980) Bull. Chem. Soc. Jpn. 53: 2815
21. Elder A, Petrucci S (1970) Inorg. Chem. 9: 19
22. Ojadi E, Selzer R, Linschitz H (1985) J. Am. Chem. Soc. 107: 7783
23. Wertz JE, Bolton JR (1972) Electron spin resonance. McGraw-Hill, New York
24. Vining JW, Caspar JV, Meyer TJ (1985) J. Phys. Chem. 89: 1095
25. Okura I, Kaji S, Aono S, Yamada A (1985) J. Mol. Catal. 32: 119
26. Fuoss RM (1958) J. Am. Chem. Soc. 80: 5059
27. Eigen M (1954) Z. Phys. Chem. N. F. 1: 176
28. Haim A (1985) Comments Inorg. Chem. 4: 113
29. Gutmann V, Resch G (1982) Comments Inorg. Chem. 1: 265
30. Gutmann V (1976) Coord. Chem. Rev. 18: 235
31. Soukup RW, Schmid R (1985) J. Chem. Educ. 62: 459
32. Schmid R, Sapunov V (1982) Non-formal kinetics. Verlag Chemie, Weinheim
33. Gutmann V (1971) Chemische Funktionslehre, Springer, Vienna, New York
34. Spange S, Guehrs KH, Heublein G, Klemm E (1984) Z. Chem. 24: 154
35. Spange S, Heublein G (1985) Z. Chem. 25: 288
36. Toma HE (1979) Can. J. Chem. 57: 2079
37. Yoneda H, Sakaguchi U, Nakamura H (1987) Bull. Chem. Soc. Jpn. 60: 2283
38. Geselowitz DA, Hammershoi A, Taube H (1987) Inorg. Chem. 26: 1842
39. Osvath P, Lappin AG (1986) J. Chem. Soc. Chem. Commun. 1056
40. Osvath P, Lappin AG (1987) Inorg. Chem. 26: 195
41. Marusak RA, Osvath P, Kemper M, Lappin AG (1989) Inorg. Chem. 28: 1542
42. Kondo S, Sasaki Y, Saito K (1981) Inorg. Chem. 20: 429
43. Ballardini R, Gandolfi MT, Balzani V, Scandola F (1987) Gazz. Chim. Ital. 117: 769
44. Rybachenko VI, Gershikov AG, Panchenko BV, Chotiy CJ, Titov EV (1987) J. Mol. Struct. 157: 321
45. Tachiyashiki S, Yamatera H (1986) Inorg. Chem. 25: 3209
46. Langford CH, Sasseville RLP (1981) Can. J. Chem. 59: 647
47. v. Smoluchowski M (1916) Physik. Z. 17: 557
48. Debye P (1942) Trans. Electrochem. Soc. 82: 265
49. Sutin N (1983) Prog. Inorg. Chem. 30: 441
50. Neumann MG, Spirandeli M, Gessner F (1986) J. Photochem. 32: 379
51. Neumann MG, Matthews JI, Braslavsky SE (1984) Photochem. Photobiol. 39: 31
52. Bagamura T, Sakai K (1988) Ber. Bunsenges. Phys. Chem. 92: 707
53. Moody GJ, Owusu RK, Slawin AMZ, Spencer N, Stoddart JF, Thomas JDR, Williams DJ (1987) Angew. Chem. 99: 939
54. Linhard M (1944) Z. Elektrochem. 50: 224
55. Vogler A, Osman AH, Kunkely H (1985) Coord. Chem. Rev. 64: 159
56. Vogler A, Kisslinger J (1982) Angew. Chem. 94: 64

57. Stich G (1989) Statische spektrale Sensibilisierung ausgewählter Cobalt(III)komplexe durch Ionenpaarbildung. Thesis, Karl-Marx-Universität, Leipzig
58. Yokoyama H, Yamatera H (1971) Bull. Chem. Soc. Jpn. 44: 1725
59. Hennig H, Billing R, Benedix R (1986) Monatsh. Chem. 117: 51
60. v. Kiss A, v. Czegedly D (1938) Z. anorg. allg. Chem. 239: 27
61. Rehorek D, Schmidt D, Hennig H (1980) Z. Chem. 20: 223
62. Miralles AJ, Szecsy AP, Haim A (1982) Inorg. Chem. 21: 697
63. Hennig H, Benedix R, Billing R (1986) J. Prakt. Chem. 328: 829
64. Larsson R (1967) Acta Chem. Scand. 21: 257
65. Schmidtke HH (1963) Z. Phys. Chem. N. F. 38: 170
66. Nakashima M, Kida S (1982) Bull. Chem. Soc. Jpn. 55: 809
67. Walther D (1983) Beiträge zur spektralen Sensibilisierung von Cobalt(III)komplexen, Thesis, Karl-Marx-Universität Leipzig
68. Figueiredo P, Pina F (1988) J. Photochem. Photobiol. A 44: 57
69. Pina F, Ciano M, Mulazzani QG, Venturi M, Balzani V, Moggi L (1984) Scient. Papers Inst. Phys. Chem. Res. 78: 166
70. Sotomayor J, Costa JC, Mulazzani QG, Pina F (1989) J. Photochem. Photobiol. A 49: 195
71. Sugimoto H, Hataoka H, Mori M (1982) J. Chem. Soc. Chem. Commun. 1301
72. Ramasami T, Endicott JF (1984) Inorg. Chem. 23: 3324
73. Clark SF, Watts RJ, Dubois DL, Connolly JS, Smart JC (1985) Coord. Chem. Rev. 64: 273
74. Bockman TM, Kochi JK (1988) J. Am. Chem. Soc. 110: 1294
75. Kunkely H, Vogler A (1989) J. Organomet. Chem. 372: 29
76. Lee KY, Kochi JK (1989) Inorg. Chem. 28: 567
77. Vogler A, Kunkely H (1988) Organometallics 7: 1449
78. Curtis JC, Meyer TJ (1978) J. Am. Chem. Soc. 100: 6284
79. Toma HE (1980) J. Chem. Soc. Dalton Trans. 471
80. Vogler A, Kunkely H (1988) Inorg. Chim. Acta 144: 149
81. Waysbort D, Evenor M, Navon G (1975) Inorg. Chem. 14: 514
82. Armor JN, Taube H (1971) Inorg. Chem. 10: 1570
83. Curtis JC, Meyer TJ (1982) Inorg. Chem. 21: 1562
84. Sexton DA, Curtis JC, Cohen H, Ford PC (1984) Inorg. Chem. 23: 49
85. Elsbernd H, Beattie JK (1968) Inorg. Chem. 7: 2468
86. Vogler A, Kisslinger J (1982) J. Am. Chem. Soc. 104: 2311
87. Vogler A, Kunkely H (1988) Inorg. Chim. Acta 150: 3
88. Vogler A, Kunkely H (1987) Inorg. Chem. 26: 1819
89. Sabbatini N, Bonazzi A, Ciano M, Balzani V (1984) J. Am. Chem. Soc. 106: 4055
90. Sabbatini N, Balzani V (1985) J. Less-Common Met. 112: 381
91. Vogler A, Kunkely H (1986) J. Chem. Soc. Chem. Commun. 1616
92. Baker WA, Phillips MG (1965) Inorg. Chem. 4: 915
93. Flamini A, Poli N (1988) Inorg. Chim. Acta 150: 149
94. Ward MD, Calabrese JC (1989) Organometallics 8: 593
95. Feldman M, Winstein S (1962) Tetrahedron Lett. 19: 853
96. Harmon KM, Cummings FE, Davis DA, Diestler JD (1962) J. Am. Chem. Soc. 84: 120
97. Kosower EM (1964) J. Org. Chem. 29: 956
98. Komatsu K, Aonuma S, Takeuchi K, Okamoto K (1989) J. Org. Chem. 54: 2038
99. Feigel M, Kessler H, Walther A (1982) Chem. Ber. 111: 2947
100. Le Goff E, La Count RB (1963) J. Am. Chem. Soc. 85: 1354
101. Beaumont TG, Davis KMC (1968) J. Chem. Soc. B: 1010
102. Billing R, Rehorek D, Salvetter J, Hennig H (1988) Z. anorg. allg. Chem. 557: 234
103. Levashova TW, Gurski ME, Ptitsyna OA, Reutov OA (1972) Izv. Akad. Nauk SSSR, Ser. Khim. 1280
104. Devoe RJ, Sahyun MRV (1987) Can. J. Chem. 65: 2342
105. Vogler A (1988) Abstracts, VIth SOPTROCC, Smolenice p 175
106. Kosower EM (1965) Progr Phys Org. Chem. 3: 92
107. Mackay RA, Landolph JR, Poziomek EJ (1971) J. Am. Chem. Soc. 93: 5026

108. Briegleb G, Jung W, Herre W (1963) Z. Phys. Chem. N. F. 38: 253
109. Kosower EM (1958) J. Am. Chem. Soc. 80: 3253
110. Toma HE, Chagas HC (1978) An. Acad. Brasil Cienc. 50: 487
111. Kjaer AM, Kristjansson I, Ulstrup J (1986) J. Electroanal. Chem. 204: 45
112. Curtis JC, Sullivan BP, Meyer TJ (1980) Inorg. Chem. 19: 3833
113. Lahner S, Wakatsuki Y, Kisch H (1987) Chem. Ber. 120: 1011
114. Fernandez A, Görner H, Kisch H (1985) Chem. Ber. 119: 1936
115. Chiorbolo C, Scandola F, Dümler W, Kisch H (1989) Abstracts, Int. Symp. Photochem. Synth. Catal., Ferrara p 127
116. Macfarlane AJ, Williams RJP (1969) J. Chem. Soc. A 1517
117. Ebbesen TW, Ferraudi G (1983) J. Phys. Chem. 87: 3717
118. Sullivan BP, Dressick WJ, Meyer TJ (1982) J. Phys. Chem. 86: 1473
119. Barnett JR, Hopkins AS, Ledwith A (1973) J. Chem. Soc. Perkin Trans. II: 80
120. Deronzier A, Esposito F (1983) Nouv. J. Chim. 7: 15
121. Prasad DR, Hoffman MZ, Mulazzani QG, Rodgers MAJ (1986) J. Am. Chem. Soc. 108: 5135
122. Doizi D, Mialocq JC (1987) J. Phys. Chem. 91: 3524
123. Prasad DR, Hoffman MZ (1986) J. Chem. Soc. Faraday Trans. 2 82: 2275
124. Deronzier A (1982) J. Chem. Soc. Chem. Commun. 329
125. Nagamura T, Sakai K (1986) J. Chem. Soc. Chem. Commun. 810
126. Calderazzo F, Pampaloni G, Lanfranchi M, Pelizzi G (1985) J. Organomet. Chem. 296: 1
127. Knoll H, Billing R, Hennig H, Stufkens DJ (1990) Inorg. Chem., in press
128. Badilescu S, Manu L, Balaban AT (1979) Rev. Roum. Chim. 24: 947
129. Tamamura T, Yokohama M, Kusabayashi S, Mikawa H (1974) Bull. Chem. Soc. Jpn. 47: 442
130. Maeda K, Miyahara T, Mishima T, Yamada S, Sano Y (1984) J. Chem. Soc. Perkin Trans. II: 441
131. Blandamer MJ, Gough TE, Griffiths TR, Symons, MCR (1963) J. Chem. Phys. 38: 1034
132. Nickol SL, Kampmeier JA (1973) J. Am. Chem. Soc. 95: 1908
133. Walsh TD, Long RD (1967) J. Am. Chem. Soc. 89: 3943
134. Jones II G, Goswami K (1986) J. Phys. Chem. 90: 5414
135. Pina F, Maestri M (1988) Inorg. Chim. Acta 142: 223
136. Marsagishvili TA, Khoshtariya DE (1987) Khim. Fiz. 6: 1511
137. Cannon RD (1980) Electron transfer reactions, Butterworths London
138. Cannon RD (1978) Adv. Inorg. Radiochem. 21: 179
139. Brunschwig BS, Ehrenson S, Sutin N (1986) J. Phys. Chem. 90: 3657
140. Symons MCR, Trevalion PA (1962) J. Chem. Soc. 3503
141. Hush NS (1967) Prog. Inorg. Chem. 8: 391
142. Gutmann V, Gritzner G, Danksagmueller K (1976) Inorg. Chim. Acta 17: 81
143. Messina A, Gritzner G (1979) J. Electroanal. Chem. 101: 201
144. Mayer U, Gerger W, Gutmann V, Rechberger P (1980) Z. anorg. allg. Chem. 464: 200
145. Mayer U, Kotocova A, Gutmann V, Gerger W (1979) J. Electroanal. Chem. 100: 875
146. Marcus RA (1956) J. Chem. Phys. 24: 956
147. Sullivan BR, Curtis JC, Kober EM, Meyer TJ (1980) Nouv. J. Chim. 4: 643
148. Tom GM, Creutz C, Taube H (1974) J. Am. Chem. Soc. 96: 7827
149. Hennig H, Rehorek A, Thomas P (1984) Inorg. Chim. Acta 86: 41
150. Blackbourn RL, Doorn SK, Roberts JA, Hupp JT (1989) Langmuir 5: 696
151. Frank R, Rau H (1983) J. Phys. Chem. 87: 5181
152. Aono S, Okura I, Yamada A (1985) J. Phys. Chem. 89: 1593
153. Okura I, Kita T, Aono S, Kaji S (1986) Inorg. Chim. Acta 116: 53
154. Darwent JR, Mc Cubbin I, Phillips D (1982) J. Chem. Soc. Faraday Trans. 2, 78: 347
155. Neckers DC (1989) J. Photochem. Photobiol. A 47: 1
156. Walkow F, Rehorek D (1989) J. Prakt. Chem. 331: 89
157. van der Zwan G, Hynes JT (1983) Physica 121: 227
158. Creutz C, Kroger P, Matsubara T, Netzel TL, Sutin N (1979) J. Am. Chem. Soc. 101: 5442

159. Nagamura T, Sakai K (1987) Chem. Phys. Lett. 141: 553
160. Briegleb G, Herre W, Jung W, Schuster H (1965) Z. Phys. Chem. N. F. 45: 229
161. Balzani V, Scandola F (1988) In: Fox MA, Chanon M (eds) Photoinduced electron transfer, Elsevier, Amsterdam, part D, p 148
162. Adamson AW, Sporer AH (1958) J. Am. Chem. Soc. 80: 3865
163. Manfrin MF, Varani G, Moggi L, Balzani V (1969) Mol. Photochem. 1: 387
164. Howes KR, Pippin CG, Sullivan JC, Meisel D, Espenson JH (1988) Inorg. Chem. 27: 2932
165. Figuereido P, Pina F (1988) J. Photochem. Photobiol. A 44: 57
166. Pina F, Ciano M, Moggi L, Balzani V (1985) Inorg. Chem. 24: 844
167. Balzani V, Carassiti V (1970) Photochemistry of coordination compounds, Academic, London
168. Adamson AN, Sporer AH (1958) J. Inorg. Nucl. Chem. 8: 209
169. Kraut B, Vincze L, Papp S (1986) Acta Chim. Hung. 122: 203
170. Hennig H, Walther D, Thomas P (1983) Z. Chem. 23: 446
171. Pina F, Mulazzani QG, Venturi M, Ciano M, Balzani, V (1985) Inorg. Chem. 24: 848
172. Norden B (1971) Acta Chem. Scand. 25: 2776
173. Pina F, Maestri M, Ballardini R, Mulazzani QG, D'Angelantonio M, Balzani V (1986) Inorg. Chem. 25: 4249
174. Bockman TM, Kochi JK (1989) J. Am. Chem. Soc. 111: 4669
175. Kosower EM, Lindquvist L (1965) Tetrahedron Lett. 50: 4481
176. Nagamura T, Sakai K (1988) J. Chem. Soc. Faraday Trans. 1, 84: 3529
177. Ebbesen TW, Manring LE, Peters KS (1984) J. Am. Chem. Soc. 106: 7400
178. Hoffman MZ, Prasad DR, Jones II G, Malba V (1983) J. Am. Chem. Soc. 105: 6360
179. Jones II G, Zisk MB (1986) J. Org. Chem. 51: 947
180. Chiorbolo C, Scandola F, Kisch H (1988) Abstracts, XIIth IUPAC Symp. Photochem., Bologna p 480
181. Klein D, Moeller CW, Ward R (1958) J. Am. Chem. Soc. 80: 265
182. Rehorek D, Hennig H (1980) Z. Chem. 20: 420
183. Rehorek D, Hennig H (1979) Z. Chem. 19: 263
184. Langford CH, Vuik CPJ, Kane-Maguire NAP (1975) Inorg. Nucl. Chem. Lett. 11: 377
185. Vuik CPJ, Kane-Maguire NAP, Langford CH (1975) Can. J. Chem. 53: 3121
186. Rehorek D, Hennig H (1980) Z. Chem. 20: 109
187. Nagorsnik E, Rehorek D, Thomas P (1974) Z. Chem. 14: 366
188. Rehorek D, Ackermann M, Hennig H, Thomas P (1979) Z. Chem. 19: 149
189. Ahn BT, Mc Millin DR (1978) Inorg. Chem. 17: 2253
190. Ahn BT, Mc Millin DR (1981) Inorg. Chem. 20: 1427
191. Rybak W, Haim A, Netzel TL, Sutin N (1981) J. Phys. Chem. 85: 2856
192. Bolletta F, Maestri M, Moggi L, Balzani V (1974) J. Phys. Chem. 78: 1374
193. Juris A, Manfrin MF, Maestri M, Serpone N (1978) Inorg. Chem. 17: 2258
194. Demas JN, Addington JW (1976) J. Am. Chem. Soc. 98: 5800
195. Ballardini R, Gandolfi MT, Balzani V (1985) Chem. Phys. Lett. 119: 459
196. Deronzier A, Meyer TJ (1981) Inorg. Chem. 19: 2912
197. Kennelly T, Streka TC, Gafney HD (1986) J. Phys. Chem. 90: 5338
198. White HS, Becker WG, Bard AJ (1984) J. Phys. Chem. 88: 1840
199. Staffan CR, Bakac A, Espenson JH (1989) Inorg. Chem. 28: 2992
200. Schmehl RH, Whitten DG (1981) J. Phys. Chem. 85: 3473
201. Kalyanasundaram K, Graetzel M (1980) Helv. Chim. Acta 63: 478
202. Usui Y, Misawa H, Sakuragi H, Tokumaru K (1987) Bull. Chem. Soc. Jpn. 60: 1573
203. Fischer T, Schiller K (1985) J. Inf. Rec. Mat. 13: 265
204. Ebbesen TW, Previtali CM, Karatsu T, Arai T, Tokumaru K (1985) Chem. Phys. Lett. 119: 489
205. Görner H (1985) J. Phys. Chem. 89: 4112
206. Chatterjee S, Gottschalk P, Davis PD, Schuster GB (1988) J. Am. Chem. Soc. 110: 2326
207. Bird GR, Pandolfe WD, Shimizu S (1978) Photogr. Sci. Eng. 22: 122

208. Böttcher H, Epperlein J (1988) Moderne photographische Systeme, 2. Aufl., Verlag f. Grundstoffind., Leipzig
209. Calvert JG, Pitts jr, JN (1966) Photochemistry, Wiley, New York, p 471
210. Cox A, Kemp TJ, Payne DR, Pinot de Moira P (1977) J. Photogr. Sci. 25: 208
211. Becker HGO, Fanghähnel E, Schiller K (1974) Wiss. T. Techn. Hochsch. Leuna-Merseburg 16: 322
212. Becker HGO, Israel G, Oertel U, Vetter HU (1985) J. Prakt. Chem. 327: 399
213. Eberson L (1982) Adv. Phys. Org. Chem. 18: 79
214. Brede O, Mehnert R, Naumann W, Becker HGO (1980) Ber. Bunsenges. Phys. Chem. 84: 666
215. Romming C, Waerstad K (1965) J. Chem. Soc., Chem. Commun. 299
216. Becker HGO (1978) Wiss. Z. Techn. Hochsch. Leuna-Merseburg 20: 253
217. Doyle MP, Guy JK, Brown KC, Mahapatro SN, van Zyl CM, Pladziewicz JR (1987) J. Am. Chem. Soc. 109: 1536
218. Pappas SP (1985) J. Imaging Technol. 11: 146
219. Pappas SP, Pappas BC, Tilley M, Yagci Y, Schnabel W (1988) Abstracts, XIIth IUPAC Symp. Photochem., Bologna, p 519
220. Devoe RJ, Sahyun MRV, Schmidt E, Serpone N, Sharma DK (1988) Abstracts, XIIth IUPAC Symp. Photochem., Bologna p 475
221. Levit AF, Korostyshevskii IZ, Gragerov IP (1970) Zhurn. Org. Khim. 6: 1878
222. Smith GH, Olofson PM (1979) US-Pat. 4173476
223. Milazzo G, Caroli S (1978) Tables of standard electrode potentials, Wiley, New York
224. Lilie J, Shinohara N, Simic MG (1976) J. Am. Chem. Soc. 98: 6516
225. Creaser II, Sargeson AM, Zanella AW (1983) Inorg. Chem. 22: 4022
226. Billing R, Benedix R, Stich G, Hennig H (1990) Z. anorg. allg. Chem., in press
227. Loginov AV, Shagisultanova GA (1988) Abstracts, XIIth IUPAC Symp. Photochem., Bologna, p 211
228. Loginov AV, Gorbunova VV, Loginova NN, Shagisultanova GA (1988) Abstracts, 6th Symp. Photochemistry, Eisenach, p 158
229. Wilson BW, Solomon EI (1980) J. Am. Chem. Soc. 102: 4085
230. Larsson R (1969) Acta Chem. Scand. 23: 1780
231. Yuasa S, Haruta M, Saito K, Nishimura Y (1988) US-Pat. 4740449
232. Nakashima M, Kida S (1977) Bull. Chem. Soc. Jpn. 50: 857
233. Wilkes GR, Brault AT (1975) US-Pat. 595932
234. Brault AT, Daniel DS, Wilkes GR (1975) DE-Pat. 2516174
235. Dominh T (1975) DE-AS 2516188
236. Adin A, Fleming JC (1975) DE-AS 2516270
237. Fletcher GL, Przezdziecki WM, Wilson JC, Yocobucci PD, van Hanehem RC (1978) US-Pat. 4247625
238. Adin A, Wilson JC (1980) US-Pat. 4239848
239. Scandola MA, Scandola F, Carassiti V (1969) Mol. Photochem. 1: 403
240. Borden DG (1972) Photogr. Sci. Eng. 16: 300
241. Linert W (1987) Inorg. Chim. Acta 132: 81
242. Eisch JJ, Boleslawski MP, Tamao K (1989) J. Org. Chem. 54: 1627
243. Moon EW, Lee BG, Kim KJ (1988) Bull. Korean Chem. Soc. 9: 209
244. Fachinetti G, Funaioli T, Zanazzi PF (1988) J. Chem. Soc. Chem. Commun. 1100
245. Bancroft EE, Blount HN, Janzen EG (1979) J. Am. Chem. Soc. 101: 3692
246. Rehorek D, Marx J (1980) J. Prakt. Chem. 322: 872
247. Feller KH, Fassler D, Gadonas R, Krasauskas V, Pelakauskas A, Piskarskas A (1989) J. Inf. Rec. Mat. 17: 459
248. Wright RF (1987) US-Pat. 4788124
249. Gottschalk P, Schuster GB (1986) US Pat. 4772541
250. Schuster GB, Gottschalk P (1988) US-Pat. 4800149
251. Kitteridge JM, Armstrong RJ (1977) US-Pat. 4126468
252. Köhler G, Solar S, Getoff N, Holzwarth AR, Schaffner K (1985) J. Photochem. 28: 383
253. Benedix R, Billing R (to be published)

254. Billing R, Sakharova GV, Atabekyan LE, Hennig H, to be published
255. Martins LJA, Kemp TJ (1984) J. Chem. Soc. Faraday Trans. 1, 80: 2509
256. Eberson L (1987) Electron transfer reactions in organic chemistry. Springer, Berlin Heidelberg New York, p 52
257. Endicott JF (1975) In: Adamson AW, Fleischauer PD (eds) Concepts in inorganic photochemistry. Wiley, New York, p 81
258. Azuma M, Hashimoto S, Hiramoto M, Watanabe M, Sakata T (1989) J. Electroanal. Chem. 260: 441
259. Hoffman MZ (1989) Inorg. Chem. 28: 978
260. Akhtar MJ, Haim A (1988) Inorg. Chem. 27: 1608
261. Skell PS, May DD (1981) J. Am. Chem. Soc. 103: 967
262. Arzhankov SI, Poznyak AL (1974) Zhurn. Priklad. Spektr. 21: 745
263. Poznyak AL, Shagisultanova GA, Arzhankov SI (1970) Zhurn. Fiz. Khim. 44: 2391
264. Mulazzani QG, D'Angelantonio M, Venturi M, Hoffman MZ, Rodgers MAJ (1986) J. Phys. Chem. 90: 5347
265. Baezold D, Fassler D (1985) Z. Chem. 25: 145
266. Aliwi S, Bamford CH (1989) J. Photochem. Photobiol. A 47: 353
267. Dance IG, Solstad PJ (1973) J. Am. Chem. Soc. 95: 7256

Author Index Volumes 151–158

Author Index Vols. 26–50 see Vol. 50
Author Index Vols. 50–100 see Vol. 100
Author Index Vols. 101–150 see Vol. 150

The volume numbers are printed in italics